INTRODUCTION TO FIRE SCIENCE AND FIRE PROTECTION

Wiley Series in Fire Science

William K. Bare
Fundamentals of Fire Prevention

William K. Bare
Introduction to Fire Science and Fire Protection

William K. Bare

INTRODUCTION TO FIRE SCIENCE AND FIRE PROTECTION

JOHN WILEY & SONS

New York • Chichester • Brisbane • Toronto • Singapore

Library of Congress Cataloging in Publication Data:

Bare, William K

Introduction to fire science and fire protection.

(Wiley series in fire science)

Includes bibliographies and index.

1. Fire prevention. 2. Fire extinction.
3. Fire prevention—Vocational guidance. 4. Fire extinction—Vocational guidance. I. Title.

TH9145.B39 628.9'2 77-14002

ISBN 0-471-01708-6

Printed in the United States of America

20 19 18 17 16 15 14 13 12 11

To Donna:
Friend, Editor, Critic,
Companion, and Wife

PREFACE

The purpose of this textbook is to provide an overview of fire science and fire protection. It is designed as an introductory course for first-year fire science students and new fire fighters, as well as for those students who are exploring possible careers in fire science. It can also be used to supplement courses in public administration and related fields.

The book begins with an exploration of job opportunities in both the public and private fire protection field and a definitive discussion of the fire science curriculum. Each chapter examines a segment of the fire protection system and is written in a matter-of-fact manner that enables the student to become familiar with the reading objectives and new terms that are presented. The main chapter text is followed by review questions and additional references. Further, each chapter builds on past chapters until the discussion of the total fire protection system is concluded with fire fighting tactics and strategy. The final chapter is then devoted to a consideration of job applications and entrance and promotional examination procedures to assist those planning public fire protection careers.

I thank Leo E. Flynn, Triton College, and Ronald W. Johnson, Western Nevada Community College, for their reviews and comments; W. Spencer Marquis, architect, for illustrative and other material; Donna Marquis-Bare for her help in preparing the manuscript; and all those who supplied information and illustrations for the text.

December 1976 William K. Bare

CONTENTS

INTRODUCTION TO FIRE SCIENCE AND FIRE PROTECTION

INTRODUCTION

For some it began when they were very young with the memory of screaming sirens, flashing red lights, and the blurred image of a charging red chariot carrying men to unknown excitement and danger. For others, a member of the family was one and they wanted to be one too. Some were searching for a career curriculum in college and stumbled upon an interesting course description, a course that stimulated the imagination. Still others were out of a job and had been combing the "Help Wanted" ads when they found what seemed to be an interesting job opportunity. A very few began by studying engineering and found a new specialty that offered interest and challenge. The very fact that you are reading these words means that you may be the next one to join and may have found your way via one of the circumstances described above. Perhaps you have taken that first step to enter the field of fire protection and fire science as one of the new breed of guardians against fire.

Sound dramatic? Yes, but it is a profession, regardless of its form, that deals with dramatic moments and sudden tragic shifts from life to death in a matter of a few seconds for someone somewhere. It is also a profession that has its share of dull routine, petty problems, frustrations, and, in the case of the Fire Service, a different life style. It is a profession in which you may be called upon to place yourself in danger to save others; it is one of the most hazardous occupations in the United States today.

To help you and others make a decision regarding the viability of a fire protection career is a major objective of this book. The text covers the basic principles of fire protection and the various components of the fire protection system. Upon completion of this volume you will not be a fire protection expert, but you will have enough information to make a decision about whether you want to pursue a career in fire protection. At the very least you will have a better understanding of the problem presented by fire and fire protection in general.

Before we begin our journey through fire protection, let's identify our objectives. They are:

1. To introduce students who are not currently in the Fire Service to the fire protection field.
2. To provide an overview of fire science, the Fire Service, and the fire protection field.
3. To identify the components of the fire protection system, specifically the public Fire Service.
4. To identify the career opportunities in the fire protection field, their basic requirements, and functions.
5. To provide the basic information needed for entry into the Fire Service and also for preparing for qualifying examinations and promotions.

If you are still interested, let's start with "The Fire Protection System."

PART I

I THE FIRE PROTECTION SYSTEM

chapter 1

1 FIRE SCIENCE - WHAT AND WHY

Most people with whom I come into contact ask sooner or later about my profession, and I have found by experience that the term "fire science" is little known outside of fire protection circles. Therefore, it seems reasonable to begin with a road map, if you will, of the terrain of fire science and an explanation of why anyone would have to go to college to become a "fireman."

The fields of fire science and fire protection have a vocabulary all their own, just like other academic disciplines, so our question about "fireman" is in trouble to begin with; a "fireman" tends boilers and other similar types of apparatus, whereas a fire fighter is a professional fire fighting person. Furthermore, the fire fighter isn't the only professional involved with fire science and fire protection. Others include fire protection engineers, fire underwriters, insurance adjustors, fire safety supervisors, Fire Service field instructors, fire prevention analysts, and safety directors. Some of these titles apply to those who work in private industry fire protection jobs, and there are other titles used in fire departments, both public and private, which we will discuss later. All of the jobs mentioned have the same basic requirement: educational courses both in fire protection and allied fields. Some require a high school diploma for entry, whereas others require at least some college work or a degree. All require continuing education after entry to keep up with the changes within the profession.

In this chapter we will define fire science, explore its content and requirements, and describe some of the jobs available in this field. We will also identify some of the requirements for the various degrees that may be obtained in fire protection and related fields, and note where these courses may be found. In short, I am going to save you some time by sharing the distillation of the research that I conducted at the public library and elsewhere. In addition, this chapter will lay the groundwork for further discussions.

At the end of the chapter we will have accomplished the following tasks:

1. To define the function of a college fire science curriculum.
2. To outline the scope and content of a fire science two-year curriculum.
3. To describe the basic requirements for a certificate and an associate of arts degree in fire science and fire protection.
4. To note some of the locations and titles of fire protection programs offered in the United States.
5. To outline and describe a four-year bachelor of arts degree program in fire protection engineering.
6. To identify the educational requirements for the various jobs available in fire protection, as well as some of the duties entailed in those jobs, and to suggest where to get more information regarding specific job requirements.

FIRE SCIENCE DEFINED

To begin our journey in fire protection, it is necessary to define the term "fire science." The simplest way to define fire science is as follows: it is a series of courses covering the basic principles of fire protection that lead eventually to a college certificate or degree. This doesn't sound too bad for a definition of fire science; yet it doesn't really explain the reason for its existence in a college curriculum, nor does it give sufficient information as to its content.

If we examine fire science a bit more carefully and divide the term into its two words, perhaps we can arrive at a clearer definition. First, let's examine the word "fire." Fire is a straightforward word, and we can guess that in this context it refers to the chemical reaction called combustion that produces heat and light.

Next, let's consider the second half of our term, "science." Now with the word "science" we are getting somewhere. A science is a branch of organized knowledge or study that is concerned with the facts, principles, and methods used in a specific area of interest (e.g., natural science, physical science). Or, it can be considered a skill or technique acquired through training and education. If we use a particular branch of science for practical purposes, it then becomes an applied science, as opposed to a theoretical or abstract science. Further, we may say that in developing tools and methods for using a science for practical purposes we create a technology. If we use our study of fire for practical purposes in the prevention and control of fire, we have a fire science or fire technology. Fire science then can refer to a series of courses of instruction which teach the basic principles of fire and fire protection. We can speak of a systematic treatment or study of the practical applications, as opposed to a study of theory.

So, in essence, fire science or fire technology is a field of knowledge about the basic principles and methods of a profession in the applied sciences, and specifically deals with the prevention and combating of fires. The fact that a college degree may or may not be given, or that it may lead to a vocation considered by some a manual skill, does not make it any less an applied science or one that does not require college instruction and training. As with other applied sciences, certain abilities can only be learned by field experience and on-the-job training; these include manipulative skills that become second nature after practice.

Now that we have a definition of fire science in hand, what

are we going to do with it in this text? Well, we are going to survey some of the courses in the fire science curriculum and give a preview of their contents and why they are required. Let's start with a fire science curriculum in a college and locate some of the schools where these curriculums are offered.

FIRE SCIENCE CURRICULUM

Fire science is called by various names, depending upon the area; whatever the name given, the basic principles of the applied science of fire protecion are the same. To get some idea of the various names given to the same study, consider the following list of some of the fire science curriculums found throughout the United States:

Fire Protection
Fire Protection Technology
Fire Protection and Safety
Fire Science
Fire Science and Administration
Fire Science Operation and Management
Fire Science Technology
Fire Technology

All these titles of college curriculums have in common the use of the word "fire," and as explained in the defining of fire science, technology and science can have the same definition. I have left out "Fire Protection Engineering," and the reason for the omission is that it is a four-year engineering program which will be explained later in this chapter.

Most of the fire science programs are found in community or junior colleges throughout the United States. One can pursue a degree in fire science in Alaska or Hawaii. He may also find it in most of the continental states, from New York to California.

The fire science curriculum offered in most colleges requires a minimum of 30 semester units or 40 quarter units for a certificate.

In order to receive an associate degree, 60 semester units or 80 quarter units are required, which takes approximately two years with a normal student workload. Bachelor degrees in fire science are not yet as available as in other fields, but a few colleges do offer a four-year degree in fire protection engineering, fire protection technology, or fire prevention technology. If one wishes to teach in the fire science field, a combination of education and field experience and a full or part-time teaching accreditation in education (usually vocational education) are usually required. Now that we know the time it takes to get a degree, let's examine a sample curriculum in fire science required for an associate degree.

As is obvious, unless a student is employed in the Fire Service (fire department) or fire protection field, the basic course required is "Introduction to Fire Science." Other fire science courses are:

Fundamentals of Fire Prevention

Fire Service Hydraulics

Fire Company Organization and Management

Hazardous Materials

Building Construction for Fire Protection

Fire Apparatus and Equipment

Fire Investigation

Fire Protection Equipment and Systems

Fire Tactics and Strategy

Related Codes and Ordinances

Rescue Practices

Wildland (Brush and Grass) Fire Control

Fire Service Communications Systems

Records and Reports

Fire Apparatus Maintenance

In addition to these major courses, English, chemistry, physics, mathematics, and other credits are required by the college. So much for the two-year fire science program. What about a four-year program in fire protection engineering?

FIRE PROTECTION ENGINEERING

Fire protection engineering is a four- to five-year course of study that is found in few colleges as of this writing. The basic requirements for a bachelor's degree involve knowledge of higher mathematics, including some study of calculus, prior to entering the college or university. To give a broad overview of a fire protection engineering program I will use a combination of the programs offered by the University of Maryland and the Illinois Institute of Technology. Both have basically the same core courses required for any degree, and the same engineering courses that are required before specialization. The fire protection engineering courses are:

Electrical Engineering
Differential Equations (Math)
Fluid Mechanics
Thermodynamics
Computer Programming
Fire Protection Equipment and Systems
Nuclear Engineering and Hazards
Research and Fire Analysis
Process and Transportation Hazards
Building Construction Fundamentals
Statistics
Systems Analysis
Municipal Fire Protection
Fire Behaviour
Organic and Inorganic Chemistry
Industrial Safety

The above list does not include the higher mathematics required at the college level, nor the basic humanities and general education requirements set by the college for a bachelor's degree.

So far, we have discussed those educational opportunities available via college; however, there are other educational opportunities for persons in fire protection, and that is what we will explore next.

FIRE ACADEMIES AND SEMINARS

Other educational opportunities are available for those in fire departments and in fire protection at annual schools or in seminars or academy programs offered in the various states. A random selection of specific states and their offerings for other than in-service and college training programs follows:

CALIFORNIA	Annual fire academy programs in fire prevention, investigation, management, alarm, apparatus maintenance, training, chief officer candidate school, and others. These programs are presented by the State Department of Education, under the supervision of the Fire Service Training Section.
KANSAS	Annual fire school, presented by the Director of Fire Service Training at the University of Kansas.
MARYLAND	Annual program presented by the Director of the Fire and Rescue Institute at the University of Maryland.
OREGON	Annual seminars at the community colleges.
PENNSYLVANIA	State Fireman's Training School, presented by the Department of Education of the Commonwealth of Pennsylvania.

In addition to these state-sponsored programs, professional associations and groups present programs and informational materials at their meetings; these include the Society of Fire Protection

Engineers, and the various Fire Service organizations at state and national levels, such as the International Fire Chief's Association, Fire Marshals of North America, and others. Now is the time, having reviewed the educational opportunities and requirements of fire science curriculum, to discuss what kinds of jobs are available in the fire protection and fire science field. What's in it for you? What is required for a particular job? Is it worth your time and effort spent to study fire science rather than some other vocational field?

OPPORTUNITIES IN FIRE PROTECTION

As mentioned previously in this chapter, there are many titles in fire protection; now is the time to explore these titles and jobs a bit further. For our purposes we will divide the jobs into two major groups: (1) private industry, and (2) governmental (federal, state, local, and volunteer). Since most of the text is going to be devoted to governmental fire protection, we will begin with jobs in private industry. The following paragraphs contain a list of job titles, a synopsis of functions, entry requirements, and where to get more information, as well as starting salaries as of this writing.

Private Industry

Fire Research Engineer. A minimum of a master's or doctorate degree and experience in fire research and development are required. The engineer is usually employed to develop and test new fire protection devices and engineer new systems. The salaries offered depend upon education and experience. These positions are usually within large corporations that range from petrochemical companies to testing laboratories.

Fire Protection Engineer. A minimum of a bachelor's degree in fire protection engineering is required. Opportunities are found within insurance companies and other private firms. Duties range from grading fire departments for the Insurance Services Office to working for the National Fire Protection Association or for consulting firms. Some work is found in jobs with federal, state, and some local governmental agencies. Again, the salary is dependent upon experience and education. The kinds of functions that may be performed by the fire protection engineer include insurance, research and product design, hazard analysis, and fire protection equipment sales. It is a relatively new profession, and some estimates indicate that nationally the opportunities are at least 20 times more numerous than the available graduate engineers.

Insurance Underwriter. Entry requirements are usually a bachelor's degree in liberal arts or business administration. Other major fields of study related to fire protection and administration can also be used. The duty of an insurance underwriter is best summarized as that of establishing the kind of risk being insured and the appropriate rates for insurance coverage of individuals, properties, and businesses for fire, theft, and other insured losses. The beginning salary is usually at or near $9000 per year. (See page 18 for places to write for more detailed information.)

Insurance Claims Adjustor. A bachelor's degree and/or experience in the field is required. For the most part, the individual insurance company will give the new adjustor additional training. An adjustor can often specialize in one type of loss, such as fire loss. This is a job that includes investigation, interviews, public relations, and a knowledge of cost and repair estimating. A person with fire science and/or fire fighting experience can utilize his knowledge to specialize in fire loss claim adjusting. The average annual salary is approximately $11,000 per year.

Fire Protection Designer and Fire Protection Designer-Estimator. A designer or designer-estimator works with fire protection equipment companies, such as those which design and install automatic fire sprinkler systems. The job usually requires some college education and drafting skills. The designer must have an excellent working knowledge of the various types of fire protection systems, code requirements, national standards, and hydraulics. Salary is dependent upon education and skills.

Fire Protection Equipment Salesman. This position is essentially a straightforward sales job that requires the ability to work independently and a knowledge of the fire protection field and its equipment requirements. Some college education is required and a bachelor's degree preferred. Experience in the Fire Service and fire protection may be substituted for degree requirements.

There are many other professions and positions that are related either directly or indirectly to the fire protection field, such as a plant safety supervisor, plant safety inspector, and plant safety director. In almost all instances some college is required for entry or to gain a management position. Clearly, a four-year fire science major gains more options for an individual. And, certainly, an experienced fire fighter with an associate degree or higher can choose from a greater variety of jobs throughout his working career. From private industry let's move to some of the opportunities available within governmental service.

Federal, State, and Local Governmental Opportunities

Federal. The federal government employs people at all levels in the fire protection field. The opportunities range from smoke jumpers with the United States Forest Service (part of the U.S. Department of Agriculture) to fire protection engineers in the Government Services Administration, which is responsible for fire protection in all federal buildings, among many other duties and responsibilities. The Armed Services also utilize personnel for a wide range of fire protection duties, including fire fighting and fire inspection. The requirements for entry into occupations related to fire protection in the federal government are as varied as the jobs. The minimum requirement is the passing of the federal government entrance examination and then qualifying for a specific job. (Becoming a specialist in fire protection in the Armed Forces, as a uniformed member, is achieved via the normal route of enlistment, physical examination, and taking your chances on getting what you want.) Opportunities for utilizing fire protection skills in the federal government service are wide ranging and depend upon your personal preference. The kinds of jobs available include wildland (brush, grass, rangeland, forest) fire fighting, municipal fire fighting (structure fire protection), fire inspection and safety (both wildland and structure), engineering and technical duties for such federal organizations as the Department of Defense, Government Services Administration, National Park Service, U.S. Forest Service, National Bureau of Standards, and the Fire Prevention and Control Administration (part of the Department of Commerce).

Most positions in federal fire protection require a minimum of a high school education and the meeting of certain physical qualifications. Government service positions are designated by the capital letters "GS," and then the rank is given by a number. The lowest level position in government service is GS3 and the highest is a GS18. Positions in the fire protection group range from GS3 for starting fire fighter to GS14, which is a chief officer. For ranks above GS7 you must have supervisory experience. The salaries in the fire protection group range from about $6000 per year to the top of about $21,000 per year for chief. For fire protection engineers the salary range is higher.

State. The opportunities in fire protection at the state government level are restricted to wildland fire protection (forest, brush, grass, and rangelands), some structure (municipal) protection, and, if there is an office of the state fire marshal, openings for fire protection engineers and specialists in fire prevention and inspection. There are also opportunities in education, working for either the state department of education or a college. Entry is based upon meeting the individual state civil service requirements for the specific job and physical examinations. Entry salaries in fire fighting start at about $850 per month. Positions requiring special skills and/or education (e.g., deputy state fire marshal, fire service instructor or supervisor) have higher starting salaries. All positions require a minimum of a high school education, and for certain jobs some college is preferred or a degree required. Again, fire protection courses can be important for gaining supervisory or higher paying specialized fire protection positions.

Local Government (County, District, Municipal). Since most of this text will be devoted to municipal fire protection functions and requirements, we will limit our discussion here to positions other than fire fighter. Entry requirements for fire fighters, work conditions, and other information pertaining to a municipal fire department will be covered in later chapters.

Positions available in the fire protection field at local government levels involve inspection and safety. For example, many local governments require a qualified protection engineer for the top job in a building inspection department, a public safety department, and a fire prevention bureau or department. Again, the entry requirements vary, but most are the same as for those positions in private industry, with field experience and education included. For example, a professional magazine recently advertised an opening for a fire protection engineer with a fire department that required a bachelor of science degree in civil, mechanical, chemical, or electrical engineering. In addition, it required either four years of administrative experience and/or technical experience.

The attraction of working for a governmental agency under civil service is one of job security rather than high pay. Private industry and civil service positions in fire protection offer almost equal fringe benefits (e.g., health plan, paid vacation); however, civil service offers, for the most part, a better retirement plan and job security after the initial probation stage. The decision as to which type of situation is more desirable is dependent upon the requirements of the individual; either way the beginner must still learn about fire science. I believe one thing is assured: the requirement for higher education in fire science will increase rather than decrease in all areas of fire protection, and be necessary for promotional examinations in the state and local fire departments.

SUMMARY

Fire science is essentially an applied science normally taught at some community colleges and at a few four-year colleges. The curriculum has different titles in different sections of the United States; however, regardless of its title, it contains the courses which give the basic principles of fire protection and fire prevention. Fire protection curriculums vary from a two-year college course to a four-year bachelor's degree program in fire protection engineering. In short, students who are interested in a career in fire protection have a variety of paths from which to choose to gain their individual goals. However, the college curriculum is only one method of gaining knowledge relating to fire protection; many states provide additional training and education within their seminar and academy programs. In addition, there is on-the-job training in the public fire service.

A student can choose a career in fire protection either in private industry or government service. Within the insurance industry there are openings in adjusting, underwriting, and engineering. Fire protection equipment and systems companies offer other opportunities for fire research engineering and specialization. Most of the positions in private industry require at least some college education, if not a bachelor's degree.

The federal, state, and local governments offer careers in fire fighting, code enforcement, and research. A person seeking a career with a government agency will be involved in public service and can specialize within the public fire protection field. With the exception of engineering positions, college education may not be mandatory yet, but will probably be so within a very few years. At present the demands of a job in fire protection require that education be continued to help the individual keep up with the rapid changes in fire protection.

REVIEW QUESTIONS

1. In your own words, define fire science.
2. Do you think that a college level curriculum for fire science is necessary? Explain your answer.
3. What are the types of degrees offered in fire science and fire protection curriculums? List them and explain their differences.

4. Identify and list the major differences between a fire protection engineering curriculum and a fire science curriculum.
5. Are there educational opportunities other than college curriculums, available which relate to fire protection? What? Where?
6. List the kinds of jobs available in fire protection in the following areas, and explain a bit about each one's requirements:
   Private industry
   Federal government
   State government
   Local government
7. Which kind of job in fire protection appeals the most to you? List the reasons why you prefer that particular job over others in fire protection.

REFERENCES

J. G. Ferguson Publishing Company, *The Encyclopedia of Careers and Vocational Guidance,* Vol. II, 3rd Ed., Chicago, IL, 1975.

The Institute for Research, *Careers Nationwide with Fire Departments, Fire fighters, Fire Prevention,* No. 177, Chicago, IL, 1972.

U. S. Department of Labor, *Occupational Outlook Handbook,* 1974–75 Ed., Bulletin 1785, Bureau of Labor Statistics, Washington, D.C., 1975.

CCM Information Corp., (A subsidiary of Crowell Collier and Macmillan), *The Bluebook of Occupational Education,* New York, 1971.

For Information About Specific Careers

Insurance Information Institute
110 William Street
New York, NY 10038

National Association of Public Adjustors
1613 Munsey Building
Baltimore, MD 21202

chapter 2

2 THE PROBLEM IS FIRE

Now that we have defined fire science and know about some of the jobs available in fire protection, it is appropriate to find out what kind of problems we are faced with. It is time to identify and understand the workings of the enemy.

Fire is all around us every day in one form or another, but most people do not actually realize the potential of fire for destruction. There is a general lack of understanding about what fire is and how it can quickly turn from an insignificant flame into a roaring inferno beyond man's control.

Unless one has witnessed a house fully involved in flame and been forced to go inside and attempt extinguishment, words are inadequate to describe fire's might and fury when uncontrolled. The thick, smoke-filled air makes it difficult to see. The smoke and by-products of combustion irritate the eyes to the point of blurring the vision with protective tears. The heat of the fire sucks the moisture from the lungs; the lack of oxygen and the toxic gases produced by fire cause violent choking sensations in the body. The roaring and crackling of burning materials deafen a person to the point of near panic, even when the individual is thoroughly trained.

In a crowded area, such as a filled theater, the cry of "Fire!" can strike panic even though smoke or flames may not be visible at all. The very word *fire* can upset people so much that they trample each other in primal terror, for it arouses one of our basic instinctive reactions: survival, fear of burning, and uncontrolled fire. Once fire is seen as uncontrolled, it has the power to charm an untrained or unprepared person into stunned disbelief, then fear, then panic, and maybe even death.

In this chapter we will first discuss the basic chemistry of fire and how fire spreads, and then we will explore the effects of fire and its causes. At the end of this chapter we will have:

1. Identified the basic components of fire as a chemical reaction.
2. Identified the major stages of fire.
3. Identified the main factors that influence fire spread and fire behavior.
4. Identified the effects of fire on man and his environment.
5. Identified the major causes of fire in the United States.

New Terms

Combustion or oxidation reaction	*Conduction*
Incipient stage	*Convection*
Smouldering stage	*Radiation*
Flashover	*Fire cause*

WHAT IS FIRE?

This question may seem a bit of a sham, since almost all of us have known what fire is from the time we are old enough to comprehend the world around us. But rarely do we think of the complex reaction that produces fire; we think in terms of flames. Perhaps at the mention of "fire," we conjure up mental pictures of various sorts; one of them is likely to be of a cozy warm fire in a fireplace, glowing while nature's wind howls about the rafters of the house. But, for an understanding of fire protection and fire prevention, we must go beyond this image and delve into the chemical reaction that causes the visual and sensory perceptions of heat and light in a fireplace. Therefore, what is fire?

Fire is a chemical reaction that occurs when fuel, heat, and oxygen are present in the proper ratio, and in sufficient amounts to produce heat and light (energy). This chemical reaction is referred to as fire, oxidation, or combustion. All three terms refer to essentially the same reaction. Going back to our fireplace image, we can identify the three items—fuel, heat, oxygen—in the process we use to start a fire. We carefully prepare the various forms of fuel: newspaper, small shavings or splinters (kindling), and then larger pieces of wood or coal. We carefully arrange the fuel in the fireplace, perhaps without thinking, to take full advantage of the available oxygen in the air. We leave air space around the fuel and make sure the fireplace damper is open so we have a good flow of air (and to insure we don't fill the house up with smoke by having the damper obstruct the chimney). We next apply heat in some form. Quite possibly, we strike a match and apply it to the edges of the newspaper or shavings (fuel), and, if the heat is applied to the proper surface and for a sufficient amount of time, the first wisps of smoke and tiny flames begin. Then in a short period of time we have a blazing fire; we may feel the heat increase on our faces, and see the light from the fire brighten the room, even though our original source of heat (the match) has long since done its job. But why does the fire continue to burn?

The reason it continues to burn, until we fail to add more fuel to the fire, is that a chemical chain reaction occurs. Once a fire has started, it is a self-feeding, self-generating reaction, limited by the amount of fuel available, reduction of heat, insufficient oxygen, or the intervention of man in his attempts to extinguish the fire.

To extinguish or halt this chemical reaction, called "fire," we must do one of four things. These are:

1. Remove or reduce the heat, either by cooling with water (which absorbs heat energy) or by spreading the fuel apart

so the surrounding air will lower the heat by convection (which will be explained later in this chapter).

2. Cut off the supply of oxygen to the fire by creating an inert atmosphere (one that will not support combustion because of insufficient oxygen content) through using extinguishing agents, such as carbon dioxide, or placing dirt over the fire to prevent oxygen from reaching the reaction (smothering).
3. Remove the fuel by spreading it apart, thereby reducing the amount of available fuel for the reaction.
4. Use one of the various kinds of chemicals (e.g., multipurpose powder, Halon extinguishers) that will interfere with the chain reaction portion of the chemical reaction.

In essence, we reverse the process that we use to begin the fire. We take away one or more of the essential components of the reaction. Now that we have in hand the basic facts about the chemistry of fire, let's move to the stages of fire.

STAGES OF FIRE

Knowing the stages of fire will later help us understand extinguishers, the importance of fire detection equipment, and a bit more about the problem of fire. To illustrate what we mean when we talk about stages of fire we will use an imaginary living room fire.

What we are going to witness in our fire is a "typical" series of events. The reason that typical is put in quotes is that each fire is different since its reaction is dependent upon the exact conditions of the structure involved. For our purposes, we will assume that we have a living room with the usual furnishings (i.e., chairs, couches, drapes, wall-to-wall carpet). In addition, we have a plastic wastebasket beside a desk.

In this living room a fire is about to happen. One of the occupants emptied an ashtray in the wastebasket before going to bed. Unbeknownst to him, the ashtray contained a smouldering cigarette. Let's stop at this point and identify what we have in terms of fire potential.

Our fuel in this instance is the cigarette paper and tobacco, waste paper (in the wastebasket), plastic wastebasket, and living room furnishings. Our oxygen source is the normal room atmosphere. Our heat source was originally the match or lighter which was used to ignite the cigarette; now we have the smouldering cigarette as our heat source within the wastebasket. If we watch the wastebasket and don't make any effort to prevent a possible fire, we will be able to identify, beginning with the incipient stage, the four stages, or progression, of the fire as shown in Fig. 2.1.

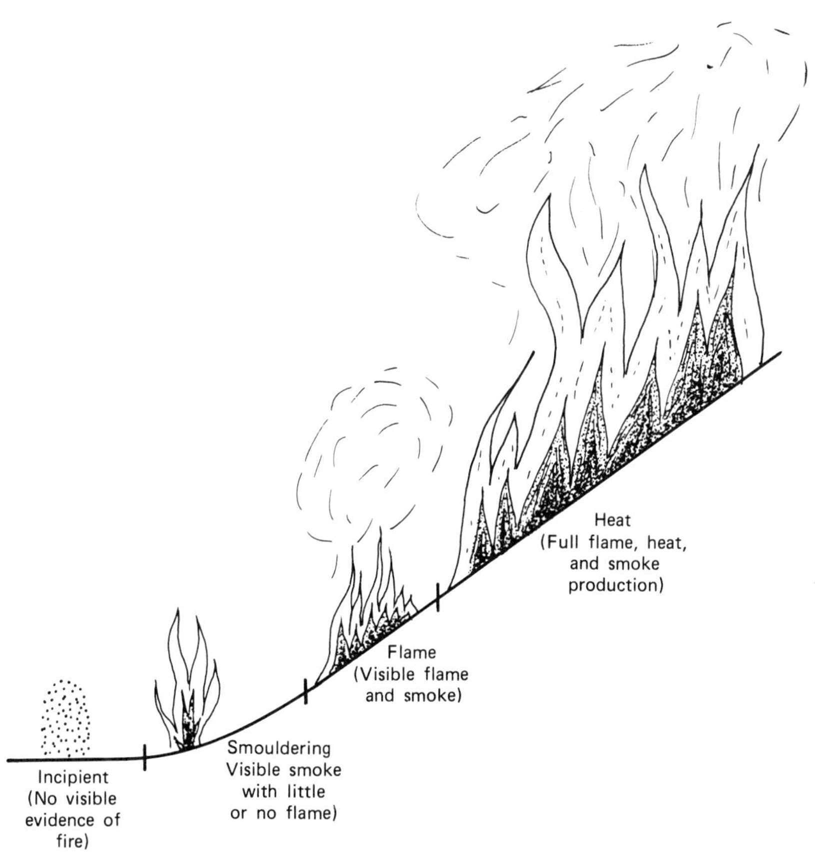

Fig. 2.1. The basic stages of fire. The duration of each stage is dependant on fuel arrangement, humidity, available oxygen, and other environmental factors.

Incipient Stage

In our living room example, the incipient stage began when the cigarette was dumped into the wastebasket. The heat given off by the smouldering cigarette began to heat the surrounding waste paper near the glowing cigarette tip. At this point, invisible products of combustion are given off, but no visible signs of fire have occurred as yet. Within a few minutes, however, we begin to detect the next stage of fire: smouldering.

Smouldering Stage

As the heat from the cigarette begins to break down the paper fibers, the paper begins to yellow and curl. Soon the heat will be great enough to reach the ignition temperature of the paper. Small tendrils of smoke begin to emerge from the wastebasket. Then we begin to see a bit of flickering light; the first flames appear, and flame production, the third stage of fire, has begun.

Flame Stage

The presence of visible flame indicates that sufficient heat, fuel, and oxygen are present to sustain the chain reaction or combustion (fire). It is self-feeding. As we watch, the flames increase in size within the wastebasket. The heat generated begins to break the plastic down into fuel vapors. These vapors ignite, and the plastic begins to burn. Smoke production increases, and we have an increased amount of toxic gases produced. Smoke indicates that the combustion reaction is not complete; therefore, along with the toxic gases, particles of unburnt fuel are being lifted into the air.

The flames continue to grow, the heat increases, the side of the desk begins to discolor from the heat, and flames begin to lick over the surface of the wall next to the wastebasket. The air temperature in the room, 70 degrees Fahrenheit at the time of the incipient stage, is beginning to climb.

The smoke has now risen to the ceiling, along with the heat, and forms a blue layer about a foot down from the ceiling. The smoke and heat is carried by air currents through open doors and through cracks around closed doors to the other parts of the house. The occupants are sleeping peacefully, unaware of the fire in the living room. Smoke will not wake them up. The fire is now building toward the final stage of intense heat and flashover.

Heat Stage

From the wastebasket the flames have climbed the wall and are licking over the ceiling. The burning, melted wastebasket began a slow creeping flame across the surface of the carpet. The desk is now burning, which in turn ignites the wall surface. The smoke production increases. The heat from the burning material ignites other fuels nearby as the flames spread outward from the point of the original fire.

The heat in the room is now rapidly climbing. The room air temperature is about 200 degrees and rising rapidly. Cool air, supplying oxygen, is drawn in along the floor from outside the house and from other rooms. It is being replaced in the other rooms by smoke-filled, superheated air.

The flames produce a roaring, crackling sound. The house begins to creak, groan, and snap as the heat increases and the air pressure builds up. The smoke level is now lower in the living room—down to within four feet of the floor. Any exits are obscured by dense, irritating smoke.

The heat continues to increase rapidly, until it is 500, 600, 700, 800 degrees, or more. The superheated air is sufficiently hot to cause ignition of portions of the room not in direct contact with open flame. At 500 degrees or above, flashover occurs (the room seems to become totally enveloped in fire in an instant). The fire now spreads through open doors to other areas of the house and through construction spaces into the attic.

The smoke with its toxic gases (carbon monoxide, hydrogen cyanide) has spread throughout the house, along with the superheated air. Anyone who tries to escape standing up will be blinded by smoke. The heated air will cook the lungs and skin. The toxic gases, such as carbon monoxide, will kill. Fire fighters will have to have their own air supply to fight the fire now, and they will have to wear protective clothing (turnouts) against the heat. At this point we will leave our occupants and their fire to examine some of the other aspects of fire behavior. Remember the stages in our fire and their effects. If we can stop a fire in its earliest stages, we can save lives and property.

FIRE BEHAVIOR

Keeping our example living room fire in mind, let's identify some of the basic aspects of fire behavior that occurred. That is, how and why does the fire spread?

The most significant factor in fire spread is the production of intense heat energy. Remember in our living room example the heat was being transmitted not only by direct contact with flame, but also by air movement. Heat was also being conducted by any metal objects in the room. Therefore, in this section, let's examine heat energy transmission. We will begin with conduction.

Conduction

A generally accepted definition of heat conduction is the transmission of heat energy through a substance without visible motion or alteration of the substance. What this definition boils down to saying is that heat applied to certain solids (steel, metal surface, wire) will eventually be transmitted throughout the solid. You can test this definition by a simple experiment.

In a safe place (over a sink), hold the end of a short piece of copper wire with your bare fingertips. At the other end of the wire apply a heat source (lighter, match). Within a short period of time, depending upon the length of the wire, the wire will become too hot to hold. The reason is that the atoms of the wire transmit heat, without visible motion, from one end of the wire to the other by conduction.

A common example of conduction is found in an electrical stove which uses electrical energy to heat the coils of the burners. (Resistance of metals and other factors come into play here but need not be discussed at this time.) Exposed steel, without a protective coating, will transmit undiminished heat energy by conduction, and thus cause ignition of any combustible materials coming in contact with it. (More on building construction will be presented in later chapters.) However, conduction is only one method of heat transfer that concerns us in fire protection. The next is convection.

Convection

In our living room fire, convection played an important role in both fire and smoke spread. Convection is defined as heat energy transmission through a substance by the motion of that substance. In other words, the heated air rose to the ceiling (because the air became lighter), and the smoke was carried with it. Cool, heavier air came in at the floor level, was heated, then rose to the ceiling. The movement of the air itself transmitted the heat energy to other surfaces; flashover occurred, and heat continued to spread throughout the house via convection.

A wall heater uses the principle of convection (and conduction) to heat our homes. Cool air is drawn in at the bottom of the heater, and is warmed by the flame and by the warm metal surfaces. The warm air rises, and cool air continues to replace the warmed air until the room temperature is the same as that set on the thermostat. Then the heater, on a signal from the temperature-sensing thermostat, shuts down until the air cools off and the thermostat signals the heater to begin the cycle again. The principle of convection in a room and in boiling water in a container is shown in Fig. 2.2. The principle is the same in both

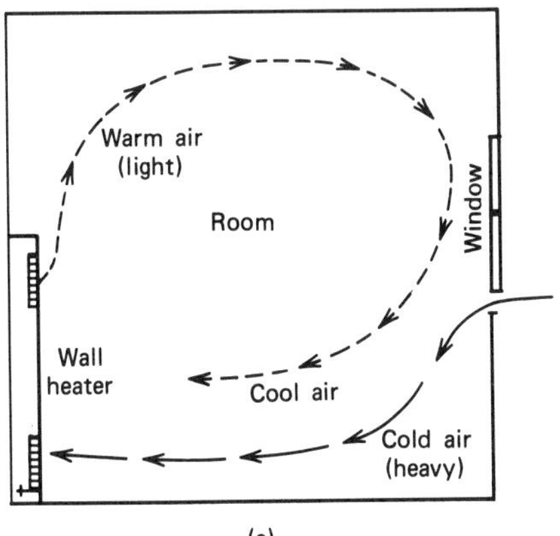

Fig. 2.2 Two examples of heat transfer via *(a)* Heat transfer by convection, and *(b)* heat transfer by conduction (burner to metal pan to water) and convection (movement of water as it is heated).

cases. Convection works in both liquid and gases, conduction in solids, and our third method of heat transfer is radiation, which acts in a manner similar to light waves.

Radiation

Radiation is the transfer of heat energy via rays traveling in a straight line from a source. The sun is our largest and most common producer of heat energy transferred through radiation.

The heat energy from the sun is transferred across space and through our atmosphere, and we feel the sun's warmth on our faces. Radiant heat can be blocked by opaque obstructions (e.g., clouds, walls, roofs), which is why, when a cloud moves between us and the sun, we feel a change in temperature. The radiant heat travels in straight lines and cannot naturally bend around a cloud. In fire protection, we rely on various methods to obstruct heat transfer by radiation in a fire. For example, a helmet and face shield are used to protect a fire fighter against radiant heat energy. If buildings are built close to property lines, the building codes prohibit windows in exterior walls in specific instances, so that possible radiant heat may not be transmitted through window glass.

Although, for purposes of discussion, we have separated the three methods of heat transfer, we must remember that all three methods occur at the same time during a fire. In both fire fighting and fire prevention, we must deal with all three methods of heat transfer in order to determine possible fire behavior in a given building or in fires outside of buildings.

A well-constructed and maintained building can help contain a fire once it has started, and with the addition of other fire protection features, such as automatic fire sprinkler systems, smoke detectors, fire dampers, and other devices, we can lessen the chances of serious fire damage and loss of life. However, if we do not have a well-constructed building and the building and fire codes have not been enforced, the fire will spread rapidly, despite the best efforts of any fire department. For example, as indicated in our sample living room fire, heat and smoke are able to penetrate throughout a building via cracks, open doors, elevator shafts, and open stairwells (stairs in multistory buildings that do not have doors that close off every floor at the stairs). Vertical shaft ways (e.g., stairs, elevators, heating and ventilating ducts) allow heat and smoke to travel as though in a chimney. In fact, these unprotected vertical openings are a major contributor to fire spread. Later, we will discuss how we can take advantage of our knowledge of fire behavior to help contain a fire and minimize damage. For now, let's move on to the causes of fire.

FIRE CAUSES

As indicated in our previous discussions, we need essentially three factors to have a fire: fuel, heat source, and oxygen. For the most part, our environment contains all three needs, so why aren't we constantly battling flames? The answer is very simple: a fire requires that all three elements (fuel, heat, and oxygen) be present in the proper relationships and forms before fire will occur. The fuel must be in a certain state (such as dust, chips, vapor); there must be enough available oxygen (we have 16 percent in our atmosphere); and the heat source must be intense enough and applied for a proper amount of time to cause ignition of the fuel and aid in the continuation of the combustion reaction. There are three things that bring these elements together every day either accidently or on purpose. These three things are men, women, and children. In all fires, except those caused by acts of nature (e.g., lightning), someone failed to do something, and a fire resulted. Each year we, as citizens of the United States, lose 12,000 people to fire deaths. Every hour, according to the President's Commission on Fire Prevention and Control in their report "America Burning," 300 fires occur, at least one person dies, and 34 of us suffer burns and injuries. Of those 34 people, many are disfigured and all are probably scarred mentally for the remainder of their lives. Fires caused by us cost an estimated $11.4 billion a year. For the most part, this fire and life loss is due to lack of public education in fire causes and fire prevention. To reinforce this point, let's examine two case histories from my own experience.

Case 1

It is an early morning weekday and the apartment is chilly. A young woman is preparing to go to work. During her preparations, she decides to stand in front of the wall heater for warmth while she puts the finishing touches on her hair.

She brushes her hair into the style she desires, and the heater provides a generous supply of heat as she prepares to put hair spray on her hair.

The metal cover on the wall heater snaps in response to the warming by the gas burner. The young woman, thinking of the work she must do that day, removes the lid from the hair spray can. She presses the aerosol can button. The hair spray is forced out of the can in a fine mist, which settles on her hair and on surrounding surfaces, including the metal heater cover. Soon she will be finished and off to work.

There is a slight pop, then a flash of light, then in milliseconds her hair is on fire. The flammable hair spray (it says flammable on the label, but who reads labels?), which keeps her hair in place, is now feeding the fire surrounding her head.

Screaming, she frantically beats at the flames with her bare hands. Fortunately, the flames die quickly, and she escapes further injury; however, she will spend time in the hospital. She will have to recover from burns on her head, neck, and hands. Her hair will eventually grow back, but it will take time.

Case 2

The young couple was tired after a long day of moving into their new apartment. The chill sea air had made the apartment feel damp and cold. One of them turned up the heater thermostat to a comfortable temperature. They had one more trip to unload the last of their boxes of clothes; then they would go out to get something to eat.

In a hurry to finish quickly, they set the cardboard boxes down wherever there was a clear space. This included setting some against the wall heater. They finished unloading, and, glad to be done, the young couple pulled the front door closed and drove off to find a place to eat in their new town.

While the couple was gone, the wall thermostat sensed the dropping room temperature and sent an electrical signal to the heater to start operating. The gas wall heater kicked on with a bit of a whoosh, then settled down to perform its normal function of heating. It performed without flaw. Not only did the heater heat the air in the room, it also heated the adjacent cardboard boxes that had been placed against it. After a short passage of time, the cardboard began to crinkle, then a tiny wisp of smoke curled from its side next to the heater. Soon, a small flame appeared, then grew. There was ample fuel.

After a relaxing dinner, the young couple returned to their apartment, looking forward to going to bed. They would unpack the next day.

They would have no need to unpack. The fire department was there to meet the young couple. Fire fighters were inside with hose lines, and other fire fighters were carrying smouldering boxes and other funishings outside. Smoke, combined with heat, flame, and water damaged or destroyed all the apartment's contents. The young couple had no insurance—everything they owned had been destroyed by the fire. They stood in shocked disbelief, she cried; he held her and stared woodenly at the smoking remains on the lawn.

In both cases human error was the basic cause. Human error provided the right combination of fuel, oxygen, and heat. The price was paid in each case. Figure 2.3 shows the leading causes of fire according to the National Fire Protection Association, but what it doesn't show is what error was committed, what was forgotten, what was caused by ignorance. However, even in situations controlled by informed experts, where we would expect everything to be taken care of, fires occur. For example, in the Apollo space program we lost three astronauts on a launch pad. Why? Everyone had prepared for the hazards in space, but forgot that a full oxygen atmosphere containing electrical equipment and ordinary fuel (paper from check lists and plans, cabin insulation) provides an ideal condition for fire. Perhaps it was an electrical short circuit or some other malfunction that generated a heat source sufficient in duration and temperature to start the fire. Whatever the cause, three astronauts strapped in a spacecraft for testing perished by fire.

Fig. 2.3 Leading causes of fire in the United States.

Heating and cooking equipment
Smoking and matches
Electrical sources
Rubbish or waste
Flammable liquids
Open flames and sparks
Lightning
Children and matches
Exposure to another fire
Incendiary (suspicious) items
Spontaneous ignition
Explosions

SUMMARY

Fire is best defined as a chemical reaction accompanied by heat and light. It may also be referred to as an oxidation reaction or a combustion reaction. In order to have a fire, we need to have fuel, oxygen, and a heat source. In addition, conditions must be just right to have the fourth factor, the chain reaction, occur if the fire is to continue without the continued application of an outside heat source. To extinguish a fire, we must remove one of the factors that caused the fire to begin.

If we examine a fire closely, we find that we can distinguish four stages in its progression. These stages are the incipient, smouldering, flame production, and heat production. The last three stages are the most destructive to the human body; the last two stages are the most damaging to a building.

Fire spread is aided and fire intensity is increased by three methods of heat transfer. These three methods are conduction, convection, and radiation. In every fire two or more of these methods of heat transfer aid the fire spread and destruction. Eventually, in a building, conduction, convection, and radiation can cause the room to become superheated. Once the air temperature is raised to the ignition point of the bulk of the fuel in the room, flashover occurs. The fire continues to spread until stopped by lack of fuel or the intervention of fire-fighting operations. Any unprotected openings within a building will increase the flame spread and influence fire behavior. A building that is properly constructed and maintained can aid in containing a fire. This containment is enhanced if the building has additional fire protection installations, such as automatic fire sprinklers.

Whatever the heat source or fuel source, most fires are caused by humans and human error. Either accidentally or on purpose (arson), approximately 300 fires per hour occur in the United States. The primary cause is people's carelessness and/or ignorance about fire and fire prevention. In addition to people causing fires directly, their use or misuse of heating and cooking equipment, smoking, and handling of other heat sources ignite property and cause life-destroying fires. Loss due to fire is $11.4 billion per year.

We must understand fire if we are to protect ourselves and others from fire's destructive effects. With a combination of public fire safety education, fire prevention, and a professional Fire Service we can do something about stemming our nation's losses and solving our fire problem.

REVIEW QUESTIONS

1. Define fire. What are the essential factors involved in fire?
2. How can a fire be extinguished? List the possible methods and describe their effect.
3. What is meant by stages of fire? How many are there? Briefly describe each stage.
4. What is dangerous about smoke? How does it affect life safety?
5. What is the major factor that aids in fire spread? How is it transmitted?
6. What construction factors influence fire spread?
7. Describe the common cause of fire. Cite your reasons for your choice.
8. List the fuel and heat sources that caused the fire to start in Case 1. How could the fire have been prevented? (Be specific in your answer.)
9. In Case 2, what is the fuel source? The heat source? List the actions that should have been taken to prevent the fire.
10. What are the three leading causes, other than people, of fire in the United States?

REFERENCES

James H. Meidl, *Flammable Hazardous Materials,* Glencoe Press, Beverly Hills, CA, 1970.

National Fire Protection Association, *Fire Protection Handbook,* Sections 1 and 2, Boston, MA, 1976.

chapter 3

3 WHO'S WHO IN FIRE PROTECTION

Most people think of the local fire department and Smoky the Bear as being the only ones involved in the fire protection field. It is not surprising, because most people's only contacts with fire protection agencies are at the local level and through television ads for preventing forest fires. (In fact, most people know how to prevent forest fires but are ignorant about protecting their homes from fire—that shows what good advertising can do for fire protection.) However, the local fire department is only the tip of the iceberg which is concerned with fire protection efforts.

In this chapter, we are going to discover who's who in fire protection and prevention; what these people do, how they affect fire protection, and how they interrelate with each other. We will be discussing private agencies, public agencies (governmental), insurance groups, professional groups, and organizations that function as testing agencies for fire protection equipment and systems. In essence, this chapter will read like a roll call of those organizations who participate, in one way or another, in finding solutions to the national and local fire problem. For ease of reference, we will divide these organizations into two main groups: private (considered part of private enterprise) and public (governmental organizations). In addition, we will devote a subsection to testing agencies.

At the end of this chapter we will have:

1. Identified the major private organizations that contribute to fire protection through control, technology, or both.
2. Identified government agencies in fire protection and their impact at the federal, state, and local level.
3. Identified the role of each agency listed, and how they contribute to the fight against fire loss.
4. Identified how those agencies listed are utilized by governmental fire protection agencies at the local level.

New Terms

Grading Schedule
Listings

We will begin our roll call of who's who in fire protection with private, nongovernmental agencies. This is an appropriate place to begin, because private concerns (insurance companies) were prime movers in gaining public fire protection and loss prevention in the United States.

PRIVATE ORGANIZATIONS

American Insurance Association (AIA)

The current nonprofit American Insurance Association was formed in 1965 when three insurance organizations combined. The three organizations were:

National Board of Fire Underwriters (NBFU), established in 1866

American Insurance Association

Association of Casualty and Surety Companies

The National Board of Fire Underwriters was instrumental in gaining changes in fire protection and building construction after the Great Chicago Fire in 1871.

The National Board of Fire Underwriters contributed to the progress of fire protection as a result of the concern of its member insurance companies about fire loss and conflagrations; fires seemed to be the rule rather than the exception in American cities during the late 1800's. To try to bring insured fire losses under control and to identify why large-loss fires in cities were occurring, the NBFU began to survey large cities. In their surveys, they noted building conditions, fire department equipment and manpower, fire fighting water supplies and other fire protection factors. When all the material was put together from the surveys, it formed the embryo of the document now known as "Standard Grading Schedule for Grading Cities and Towns of the United States with Reference to their Fire Defenses and Physical Conditions." This document is more commonly known as the "Grading Schedule." Eventually this Grading Schedule became one of the criteria by which insurance companies determine fire insurance rates for any given city or town. (More on the Grading Schedule will appear later in the chapter.) When the schedule became popular, it was utilized by regional and local rating bureaus, and the American Insurance Association carried on grading of larger cities. With the formation of the nationwide Insurance Services Offices, AIA stopped grading larger cities, and now almost all the grading is accomplished by the engineers of the Insurance Services Office. However, municipal grading was only one function of the American Insurance Association.

Although the bulk of the work and services now performed by AIA is designed for the insurance industries, the AIA provides aid, information and the results of research both to the public and the Fire Service. Among items of interest to the Fire Service, provided by AIA are:

National Building Code

Fire Prevention Code

Special Interest Bulletins (produced by the Engineering and Safety Department of AIA)

Building Code Standards for Heat Producing Appliances

In addition to the above items, the AIA also publishes educational and informational materials, such as *Recommended Methods of Laying out Fire Limits,* which serves as a basis for delineating fire zones within cities. (More about fire zones will be presented in later chapters.) All of the publications listed are available from the AIA's office at 85 John Street, New York, NY 10038. As times have changed, so has the AIA. At one time the American Insurance Association provided a respected fire and arson investigation unit; it was discontinued circa 1970, close to the time that grading of cities was relegated to the Insurance Services Office. Recent changes in the fire picture, with arson and bombings on the increase, may cause the insurance industry to demand help and bring the arson unit back into being.

The AIA building and fire prevention codes and standards are nationally recognized and respected, and its Special Interest Bulletins (there are over 300 of them), covering a multitude of subjects, should be obtained and used for reference by anyone interested in fire protection.

From the American Insurance Association, let's move on to an organization mentioned before, the Insurance Services Office.

Insurance Services Office (ISO)

The Insurance Services Office performs many functions but the one that is of primary interest to fire protection personnel and the Fire Service is that of its grading and evaluation of municipal fire defenses. The Insurance Services Office, when it was formed in 1971, took over the grading of the majority of cities from both the AIA and the regional and local rating bureaus. The Insurance Services Office is now active in the 50 states and United States territories, and is the prime rating organization or an advisory body in all but a few of the states. The states that have retained independent rating bureaus are:

District of Columbia
Hawaii
Idaho
Louisiana
Mississippi
North Carolina
Texas
Virginia
Washington

In the states listed, the ISO is active but more in an advisory capacity than in actual rating work. In the balance of the United States, the Insurance Services Office is the prime grading agency, and insurance companies base their charges for fire insurance on the rating classification of each city or town. (The rating classification is not the only factor in setting rates, but it is an important factor.)

Since this grading of fire defenses by the Insurance Services Office is of great importance to public officials and members of the Fire Service, I think it is important to explore this grading business a little bit further. After all, the grading and survey work accomplished by the ISO engineers of our cities and towns is about the only tool we have currently for comparing fire protection in various regions of the United States. If we knew the fire protection classification of the city that we live in, we could tell how well the city and the fire chief are doing in trying to provide adequate fire protection. It wouldn't give us the whole picutre, but it would give us a measurement to compare with other cities in our area.

Before we can understand what our local grading classification means, we must first understand how a grading is conducted, what the engineers look for, how the classification is determined, and what the final classification means. So let's examine a simplified version of the grading process for any city or town. First, field surveys are conducted by the ISO field engineering team in the city. These men examine and grade such things as water supply; fire department equipment, operations, manpower, training; fire prevention; fire communications; building code and code enforcement; building conditions; conflagration potential; and records and reports. Each of the areas noted are given a maximum number of potential deficiency points. For example water supply is given 1950 points with a value of 39 percent of the total maximum points (i.e., 5000). (See Table 3.1.) When the survey crew finds that conditions in any of the general categories are less adequate than the ideal of the Grading Schedule, the appropriate number of points are assigned up to the total number of deficiency points in that category. In other words, if the water supply in a given community is inadequate, that is, the supply is not up to standard, points are assigned up to the maximum of 1950 for water supply. At the end of the grading survey, the engineers from the Insurance Services Office review their findings, the points assigned are added up in each category,

TABLE 3.1 Relative Values and Maximum Deficiency Points

Feature [a]	*Percent*	*Points*
Water supply	39	1950
Fire department [b]	39	1950
Fire service communications	9	450
Fire safety control [b]	13	650
	—	——
	100	5000

[a] The data are from the "Municipal Grading Schedule" used by the Insurance Services Offices. (The reader should check the most current schedule published by ISO for possible changes.) Copyright, Insurance Services Office, 1973.

[b] "Fire department" includes items such as training, equipment, and manpower. "Fire safety control" includes building conditions, fire prevention program, building code, and enforcement.

and then a preliminary classification is assigned. At that point a conference with the fire chief and other city management personnel is arranged, and a representative from the ISO discusses their findings and offers suggestions for improvements, as well as explaining why deficiency points were assessed. A city and its fire department, armed with the grading findings, may then determine if they can improve the fire defense grading within a reasonable amount of time and before the final grading is assigned by ISO. Once the final grading is determined by the Insurance Services Office, the total points assessed against the city's fire defense are compared with the classification contained in the "Relative Grading of Municipalities in Fire Defenses and Physical Conditions." (Table 3.2). For example, if a community had a total of 1200 deficiency points assessed against it in the Grading Schedule, it would fall in the 1001–1500 point category and would be rated as a Third Class or Class 3 city for its fire defenses. In cities with classifications higher in the numerical order, the citizens pay higher fire insurance rates, especially from Class 5 through 10.

At this point in our discussion about fire defense grading, please remember that the Insurance Services Office does not set rates; they provide an assessment of fire defenses, according to the Grading Schedule, upon which insurance companies base

TABLE 3.2 Relative Grading of Municipalities in Fire Defenses and Physical Conditions [a]

Points of Deficiency	*Relative Class of Municipality*
0–500	First (Class 1)
501–1000	Second (Class 2)
1000–1500	Third (Class 3)
1501–2000	Fourth (Class 4)
2001–2500	Fifth (Class 5)
2501–3000	Sixth (Class 6)
3001–3500	Seventh (Class 7)
3501–4000	Eighth (Class 8)
4001–4500	Ninth (Class 9)
More than 4500	Tenth (Class 10)

[a] After a survey by the Insurance Services Office. (The system is subject to periodic change; the reader may want to refer to the most recent grading schedule.) Copyright, Insurance Services Office, 1973.

their own rates (consistent with state insurance regulations). Also, I have simplified the grading process in the example for purposes of illustration; grading is a complex engineering function which utilizes formulas, percentages, and deviations to evaluate the specific points of the Grading Schedule. Nonetheless, the impact of the Insurance Services Office on the public, through insurance rates and monies spent for fire protection at the local level, can be seen.

To this point we have discussed two of the private organizations dealing with fire protection (and insurance); there are others in the nation. However, for our purposes this will be sufficient, with the exception of one other insurance related organization, Factory Mutual, which will be discussed under "Private Testing Agencies". So from this point, let's move to one of the oldest and largest organizations for fire protection and prevention (not insurance-related), the National Fire Protection Association.

National Fire Protection Association (NFPA)

The National Fire Protection Association is one of the oldest (established in 1896) organizations involved in promoting and improving fire protection and prevention. It provides, among other things, fire protection educational materials for both the public and fire protection agencies. The "Bible" of fire protection, the *Fire Protection Handbook,* was first published by the NFPA in 1896 and has served as the most comprehensive fire protection publication available. In the *Fire Protection Handbook* you can find information on hazardous materials, building construction, fire protection systems, and myriad other subjects that must be dealt with in the fire protection profession.

Among other publications provided by the NFPA are the National Fire Codes, published in 16 volumes, which are nationally recognized standards for all areas of fire protection and prevention. The National Fire Codes are the combined work of the NFPA standards committees (members of NFPA), and one of the many standards is the National Electrical Code. The National Electrical Code is the basic standard for all electrical installations, and is usually adopted by local governments as their electrical code for building construction. In all, including the National Electrical Code, there are 234 standards within the National Fire Codes. In addition, much of the National Fire Codes have been adopted by the federal government, and serve as the basis of many federal programs relating to housing, industrial and occupational safety, and other concerns. Besides the production of the National Fire Codes and the *Fire Protection Handbook,* the NFPA is involved in promoting fire prevention through films, handout materials, and posters. The Association also publishes magazines that are available to its members, such as the *Fire Journal* (bimonthly), which contains a wealth of information on the latest in fire-fighting techniques, recent fire losses, fire prevention articles, and other material related to fire protection.

Throughout the remaining chapters you will find references to the various publications and standards of the National Fire Protection Association, which will give you some idea of the impact of the work of the NFPA. The NFPA has been and still is a strong force in the promotion of standardization of fire protection practices; their work ranges from providing fire department equipment standards to providing standards for all fire prevention codes and code enforcement. Let's leave the National Fire Protection Association for now and move on to the private testing organizations.

Private Testing Organizations

There are several private testing agencies in the United States, but for our purposes we need discuss only two of them: Underwriters Laboratories and the Factory Mutual Research and Engineering Section of the Factory Mutual group. Both organizations perform tests of products and list those which meet their standards. Their labels are recognized nationwide as indicating that the products have been tested for fire safety. Generally, those products that are controlled by the building and fire codes and have earned a UL or FM approval label are accepted as meeting the minimum standards contained within the codes. Let's now examine both these organizations a bit closer, beginning with the Underwriters Laboratories.

Underwriters Laboratories, Inc. (UL)

Underwriters Laboratories, Inc., is a not-for-profit, private organization that conducts tests on products used by all of us in our daily lives, as well as those on products designed for or having impact on fire and human safety. The UL has been in operation since 1894, and over the years has had a major impact on safety.

In order to receive a UL label, manufacturers submit their products for testing, and if the product meets the standards and tests for the specific product, it is then allowed to carry the applicable UL label. In addition, the product is then listed in one of the product directory lists of UL, such as:

Electrical Appliance and Utilization List
Electrical Construction Materials List
Hazardous Location Equipment List
Fire Protection Equipment List
Building Materials Directory
Gas and Oil Equipment List
Accident, Automotive, and Burglary Protection Equipment List
Marine Products List
Fire Resistance Index
Classified Products Index

All of the above directories and lists contain the manufacturer's name, the type of product, and the types of uses or installations that it has been tested for. These lists are published once a year, with semiannual supplements to keep the listings updated. The directories and listings are provided by UL, free of charge, to fire departments and code enforcement officials. Once a manufacturer's product meets UL's requirements and is listed, the product is subject to periodic reinspection and testing by UL. If the manufacturer fails to maintain the UL standard, the label and listing is withdrawn by the UL.

The testing and listing of products by UL precludes the necessity for each individual fire department and other code enforcement agencies concerned with fire safety and construction having to conduct their own testing programs to determine each product's compliance with the performance requirements of each code. In addition to the listings mentioned, UL provides *Standards for Safety* (Fig. 3.1), which is available for review. An individual can also get copies of the Standards and Directories by writing to UL.

Factory Mutual Laboratories, Engineering Division (FM)

Factory Mutual's testing and approval program is just one segment of the Factory Mutual system. Although FM's work is primarily designed for its members, it is a nationally-recognized testing agency with its own label. The FM enclosed by a diamond is a common sight on fire hoses, sprinkler equipment (automatic fire sprinklers), and other fire protection devices used in industry. Factory Mutual also publishes fire-loss prevention information and other educational materials.

Other Private Testing Agencies

There are other private testing agencies which fulfill roles in testing products and equipment, such as:

Underwriters Laboratories of Canada (ULC)

American Gas Association (AGA) which tests and lists gas appliances and equipment

Southwest Research Institute which tests construction materials, etc.

We now turn to the governmental testing agencies.

Fig. 3.1. A UL testing standard. The standard gives test specification data for the type of equipment/product noted. (Courtesy Underwriters Laboratories, Inc.)

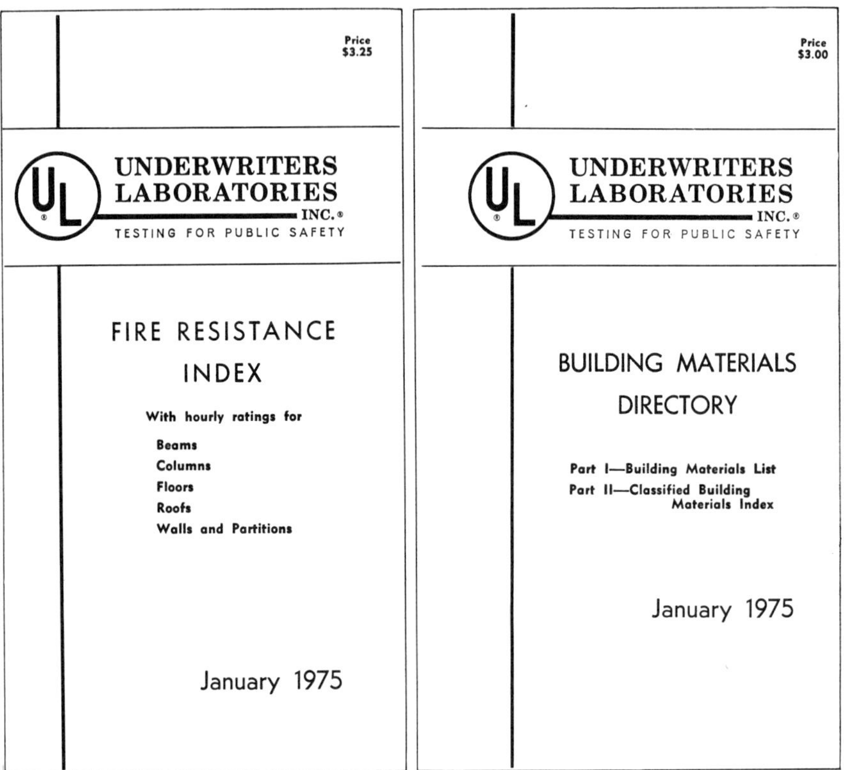

Fig. 3.2. Two examples of UL listings and directories. (Courtesy Underwriters Laboratories, Inc.)

Governmental Testing Agencies

Federal

There are many federal agencies that conduct research either directly or indirectly related to fire safety and protection. However, the agency that has the broadest base in fire protection concerns and safety research is the National Bureau of Standards (NBS) which is part of the Department of Commerce.

The National Bureau of Standards both tests products and conducts research in fire protection, aimed primarily at solving construction problems for fire safety. In addition, the National Bureau of Standards publishes minimum standards which are applied nationally, such as those pertaining to flammable fabrics (e.g., standards for fire retardant children's sleepwear and clo-

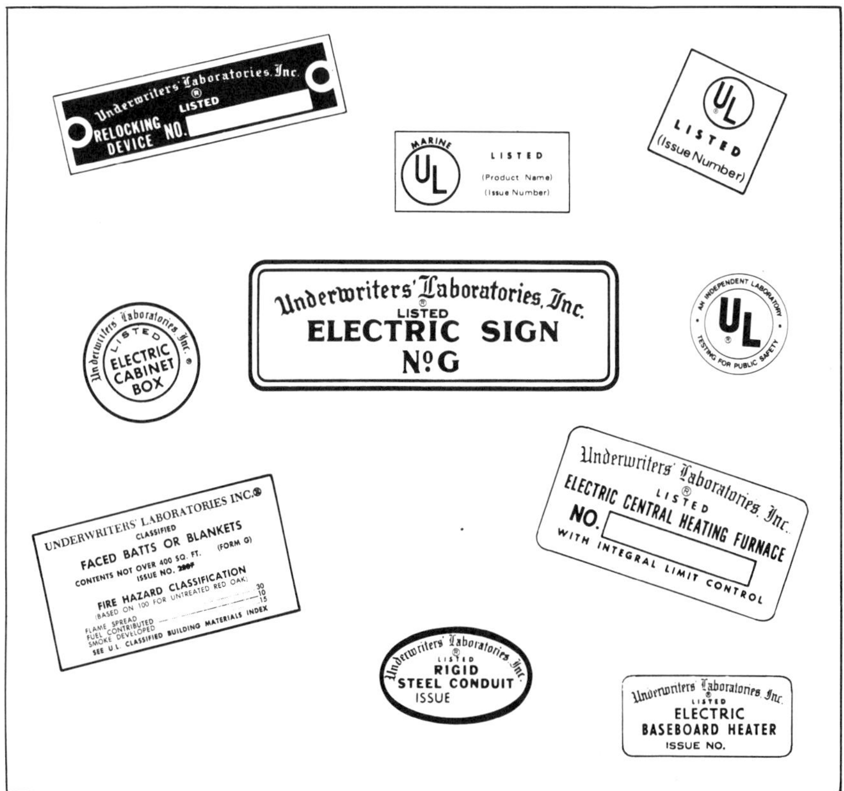

Fig. 3.3. Typical examples of UL labels found on tested and listed consumer and construction items. (Courtesy Underwriters Laboratories. Inc.)

thing). The NBS received a big boost from the passage of the Federal Fire Prevention and Control Act of 1974 which, among other things, established the NBS Fire Research Center.

Another function of the National Bureau of Standards is promoting and obtaining standardization of building and other construction requirements throughout the United States. From the federal level we move on to the testing agencies at the state level.

State

Many states support agencies that conduct research and testing; some are directly connected with the state government, while others, such as state universities and colleges, are quasi-governmental. The research and information that comes out of colleges and universities are products of individual research projects by students and faculty, or may be a part of an engineering program.

In those states where the state fire marshal's office performs a wide variety of functions, one of the functions is testing and approval of material (e.g., California State Fire Marshal's listings and testing program) and the state fire marshal publishes a list of approved materials that comply with the state's minimum fire safety regulations. These listings also include other approved testing labs, such as UL or FM.

Local

Very few local governmental agencies have the resources, personnel, or facilities either to test products or conduct fire safety research. However, two large fire departments (New York City and Los Angeles City) do perform testing and research as part of their functions. For the most part, local governments rely on the nationally-recognized testing agencies for information about products and materials.

With this, we conclude our overview of testing agencies and now move on to those governmental organizations involved in fire protection and safety. Again, only the main agencies will be discussed, and we will begin with those at the federal level and then consider those at the state level. Local fire protection organizations (fire departments) will be discussed in the next chapter.

FEDERAL GOVERNMENTAL ORGANIZATIONS

Department of Commerce

If you will recall, our first glimpse of the fire protection role of the Department of Commerce came in our discussion of testing agencies (National Bureau of Standards). Another agency of the Department of Commerce that has recently been added, via the Federal Fire Prevention and Control Act of 1974, is the National Fire Prevention and Control Administration (NFPCA).

The National Fire Prevention and Control Administration has been assigned the responsibility for improving and creating:

1. National fire loss data.
2. Educational materials for fire fighters and the Fire Service.
3. Fire Service training, including a possible National Fire Service Academy.
4. Fire protection and prevention research and development.
5. Public fire safety education and materials.
6. Funding for state and local fire protection improvements.
7. Operation of the Federal Fire Council.

At this writing, its potential effect on fire protection has not been fully realized, but the tools are there to use.

Another agency of importance to fire protection is the Maritime Administration (part of the Department of Commerce). The Maritime Administration (MARAD), among other things, sponsors research on shipboard fire protection and works with seaport cities on improving fire-fighting operations related to dockside shipboard fires.

Next on our list of federal agencies is the Department of Defense.

Department of Defense (DOD)

All of the federal military organizations (i.e., Army, Navy, Marine Corps, Air Force) are under the jurisdiction of the Department of Defense. Each of the military organizations provides fire protection and fire prevention for all the military property and facilities under its individual control. Because each one uses its own personnel for fire fighting and prevention, each branch provides training schools and information about special fire fighting problems, such as aircraft fires. As a spinoff, the Armed Services have provided a wealth of information regarding fire protection and fire fighting for all types of structures, including buildings, aircraft, and ships, and this information has been useful to civilian fire protection agencies.

Fire protection for other federal lands and property is the responsibility of two other federal agencies; they are the Department of Agriculture and the Department of the Interior.

Department of Agriculture (DOA)

The Department of Agriculture, and specifically the U.S. Forest Service, is responsible for fire protection and prevention in all those lands under the jurisdiction of USFS, that is, the national forests. In addition, the U.S. Forest Service conducts research on wood products and forest fire control and aids in promoting rural fire prevention and protection.

Department of the Interior (DOI)

The Department of the Interior is responsible for the lands in national parks and monuments, those areas under the Bureau of Land Management (BLM), and those under the jurisdiction of the Bureau of Indian Affairs (BIA). In addition, the Department of the Interior provides fire protection for private and state lands which are under contract to the Bureau of Land Management. Mine safety is the responsibility of the Bureau of Mines, which is also part of the Department of the Interior.

The DOI's Mining Enforcement and Safety Administration, along with the Bureau of Mines, provides enforcement and training for mine safety, as well as tests and approval of equipment used in mines. (Some of the equipment used in mine safety is also used in fire fighting, such as the self-contained breathing apparatus). The equipment tested and approved by the Department of Interior's mining agencies includes electrical items, breathing, and fire extinguishing equipment, systems, and apparatus.

Other Federal Agencies

Other federal agencies with some impact on fire protection are:

Department of Labor (DOL) which has overall jurisdiction for the enforcement of the Occupational Safety and Health regulations (OSHA).

Department of Transportation (DOT) which regulates the transportation and marking of hazardous materials.

Defense Civil Preparedness Agency (DCPA)—the newest name for the old Civil Defense Agency—which provides resources for surplus fire equipment, develops standards for fire equipment, and aids communities in developing a disaster plan. It has also provided information via its research programs.

Department of Health, Education, and Welfare (HEW) which has the resources, under the Emergency Medical Services Act of 1973, to provide federal funds to local communities and fire departments wishing to provide paramedic service. HEW also has been promoting, as a part of the Emergency Medical Services program, the use nationally of a single emergency number: the 911 emergency phone number.

This concludes our overview of the federal agencies concerned with fire protection. We will next explore the state government agencies.

STATE GOVERNMENT ORGANIZATIONS

Forestry Departments

Fire protection for those lands within a state that are not covered by the federal government nor by local (county or municipal fire district) fire departments is the responsibility of the state. The states may contract for fire protection of wildlands (e.g., range, forests, parks) or, as is the case in almost half of the 50 states, assign the responsibility to a department within the state government. For example, in California, the Department of Natural Resources is the main state agency for fire protection. The specific agency within the Department of Natural Resources is the California Division of Forestry (CDF), which is headed by the state forester. The Division of Forestry not only takes care of rural and wildland fire protection; it also provides investigation, prevention, inspection, public education, and in some instances fire protection of municipalities. For example, the California Division of Forestry provides fire protection, using the same equipment, for buildings and some urban areas.

Another state agency involved in fire protection and prevention is the office of the state fire marshal.

State Fire Marshal

Almost all 50 states now have a state fire marshal's office in one form or another. The functions of the state fire marshal vary with the state; they may include:

Writing and administering state fire safety regulations

Fire investigation

Enforcement coordination of fire safety laws

Fire prevention in state owned facilities

Approval and testing services for products, materials, and equipment for fire safety

In almost all instances, the local fire chief is responsible for enforcement of state fire safety regulations at the local level, but he may have assistance available from the state fire marshal's office.

State Laws Affecting Fire Protection

State laws and statutes form the basis for the organization for fire protection and authority for enforcement at the local level. Other state laws and codes cover fire fighters' retirement, vacation, collective bargaining rights, and minimum wages and hours to name a few topics. Also, some states provide tests for fire department entry and training, either through local colleges and universities or through a state fire academy training program.

Statewide disaster planning for coordination and equipment (e.g., shortwave radio networks) is another function of state government. For example, New York, California, and Oregon (as well as some other states) have a statewide plan for mobilizing and coordinating fire departments and other emergency services in the event of a disaster (e.g., earthquake, flood, civil disorder). In these states, the governor or his assigned representative can activate the plan, and fire equipment can be moved to the disaster without regard to local (municipal, county) boundaries.

Whatever the specific names of state agencies within state governments, the state is still intimately involved in both statewide and local fire protection. In essence, all of the local authority for enforcing fire prevention regulations, providing fire investigation, organizing fire protection, taxing, and regulating fire fighters' working conditions is controlled by state statutes and codes.

Now we turn to our last group of governmental fire protection organizations: local agencies.

LOCAL GOVERNMENT ORGANIZATIONS

Organizations for local fire protection (county, municipal) have various names, but they still provide the same services for their citizens. Countywide fire departments under the jurisdiction of a board of supervisors may provide both rural and urban fire protection. They may, by agreement with each city, provide fire protection services for cities and towns within the county.

Another type of local fire protection is organized by individual fire districts within the county that provide fire protection in those areas not covered by the federal, state, or city fire protection agencies. Each fire district has an elected board of directors or fire commissioners who are responsible to the citizens and taxpayers of the district and who make policy decisions regarding the chief of the department (his appointment or removal), budgeting, and, in essence, determine the operation of the fire district. The commissioner and/or directors are usually responsible to the county board of supervisors, and must use county legislative actions to gain taxes for fire protection and adoption of fire codes.

Fire prevention services may be performed by a county, either through a special office of the county fire marshal or via the use of its fire department or the fire districts. Where a county fire marshal organization is present, the county fire marshal is usually appointed with the approval of the board of supervisors. The county fire marshal is then responsible for all fire prevention within the unincorporated areas of the county (excluding cities and towns with their own prevention programs) and conducts his pogram in a manner similar to that of any fire department's prevention program (more on fire prevention programs will be presented in Chapter 7).

Within cities (towns, townships, boroughs), fire protection is provided by a municipal fire department. The chief of the department is charged with providing fire protection within the political limits (city limits) of the specific city and serves at the pleasure of the city government (manager-council, commission, mayor-council). The municipal fire department obtains its money for operating expenses from the same general tax fund as other city departments. Since the balance of the text is devoted to local fire protection, we will end our discussion of local fire protection agencies at this point; in the next chapter we will discuss a local fire protection organization in greater detail.

SUMMARY

Private, nongovernmental agencies have been involved in fire protection for many years, and in addition to serving their main commercial interest, they have contributed valuable knowledge to the art of fire protection in the United States. Many of these private organizations have been primarily insurance companies that perform many tasks relating to fire protection.

The National Board of Fire Underwriters, for example was instrumental in improving the quality of fire protection in the United States and contributed technical information regarding fire protection problems. With the merger of the National Board of Fire Underwriters, the American Insurance Association, and the Association of Casualty and Surety companies under the aegis of the American Insurance Association, the tradition of involvement in fire protection education was continued. For example, the American Insurance Association publishes two national codes (the *National Building Code* and the *Fire Prevention Code)*, as well as continuing the former NBFU Special Interest Bulletins. One aspect of both the AIA and the old NBFU has changed: they no longer conduct grading of municipalities for fire protection. This is now the province, in most states, of the Insurance Services Office.

The Insurance Services Office (ISO) is the primary agency that conducts engineering surveys relating to fire protection of cities and towns. Using the "Grading Schedule," as a standard and based on field survey, a city is assigned a grading classification. This classification is then used by insurance companies as one of the basic factors for determining the fire insurance rates of a given city or town. In addition, the grading of fire defenses helps chiefs of fire departments and city administrators determine what improvements are needed to provide better fire protection for the citizens of their communities. However, it must be remembered that in some instances (e.g., fire fighting manpower requirements) the Grading Schedule is a bit idealistic in today's economy. Another important private organization involved in fire protection and fire safety is the National Fire Protection Association.

The National Fire Protection Association (NFPA) has, since 1896, produced the *Fire Protection Handbook* and many other codes and standards that are recognized nationally. The NFPA is made up of members of both private industry and public fire protection that have contributed to and promoted fire safety, fire safety education, and fire protection standardization. The standards (contained in the National Fire Codes) are developed by NFPA members, and these standards serve as a basis for fire protection equipment and systems, code regulations, and code enforcement at all levels of government (federal, state, and local).

Private testing agencies, such as Underwriters Laboratories and Factory Mutual, play a vital role in fire protection. These and other organizations test, approve, and list a variety of products, including construction materials and household appliances, for safety. Products submitted by manufacturers to these agencies are tested according to predetermined standards; if they meet the standards, the product is then given approval and included in the listings of the particular testing agency. These approved product listings are then used by code enforcement officials, construction and fire protection engineers, architects, and others involved in fire protection. If a product is listed, approved, and installed according to the testing specifications, it is usually deemed as meeting the minimum code (e.g., building code, fire prevention code) requirements. There are also concerns that test and list specific products, such as the American Gas Association (AGA). The federal government has a testing agency of its own.

The best known federal testing agency is the National Bureau of Standards (NBS). As part of the Department of Commerce, the NBS has been involved in fire safety research and the promotion of national standards, including codes. In addition to the federal government, many states and some local governmental agencies, such as the California State Fire Marshal's Office, perform testing services.

Besides testing agencies such as the National Bureau of Standards, other governmental agencies at all levels (federal, state, local) contribute to fire protection. Federal governmental agencies range from the Department of Commerce (which includes the National Fire Prevention and Control Administration) to the Department of Transportation, which regulates interstate transportation of hazardous materials and uses hazardous materials marking and warning systems.

At the state level, we find the state fire marshal's office, which aids in enforcement and administration of minimum state fire safety standards, and, in certain states, the forestry department (e.g., California Division of Forestry). A forestry department, or its equivalent in state government, is responsible primarily for brush, forest, and grassland fire protection outside of city and town boundaries and in unincorporated areas (excluding national forests and parks, which are protected, respectively, by the Department of Agriculture's U.S. Forest Service and the Department of the Interior's National Park Service). We find that state regulations affect fire fighters' retirement, work hours, and work conditions; these state regulations also form the basis for local code enforcement. Also, many states have statewide disaster plans that coordinate all fire protection, regardless of political boundaries, in the event of a natural or man-made disaster.

Local fire protection in any given area is the responsibility of the local government agency, be it county, fire district, or city. Regardless of the type of government organization (manager-council, commission, mayor-council), the chief of the fire department is directly responsible to the local government and is dependent upon the administration and tax base for operating funds. In addition, all authority for code enforcement comes from the state to the local government agency, and then to the chief of the department as the local government's representative.

In short, many agencies beside the local fire department are involved in combating fire. All of the organizations concerned (public or private) contribute valuable information, which helps a fire fighter become a professional and indirectly increases the safety of every citizen.

REVIEW QUESTIONS

1. List the private organizations discussed in this chapter and under each outline their contributions to fire protection.
2. What functions do the testing agencies perform?
3. What is a listing? How is it used in connection with fire protection?
4. Under what federal agency does the National Bureau of Standards exist? What are the main functions of the National Bureau of Standards relating to fire protection?
5. List the functions of the National Fire Prevention and Control Administration.
6. What government agencies (federal and state) provide rural and forest fire protection? Outline the functions and responsibilities of each agency listed.
7. Interstate transportation and marking of hazardous materials is under the jurisdiction of what agency?
8. What are the basic functions of a state fire marshal?
9. What are the different types of local fire protection agencies?

REFERENCES

International City Manager's Association, *Municipal Yearbook,* Chicago, IL, 1975.

International City Manager's Association, *Municipal Fire Administration,* Chapt. 3, Chicago, IL, 1967.

National Fire Protection Association, *Fire Protection Handbook,* Appendices A, B, and C, 14th Ed., Boston, MA, 1976.

Congressional Quarterly, Inc., *Washington Information Directory,* Washington, DC, 1975.

American Insurance Association, *Municipal Fire Protection Surveys,* Special Interest Bulletin No. 297, New York, 1967.

Insurance Services Office, *Grading Schedule for Municipal Fire Protection,* New York, 1973.

Insurance Services Office, *I.S.O. Today,* New York, 1972.

chapter 4

4 PUBLIC FIRE PROTECTION

Man's struggle against fire has evolved from each individual trying to provide his own fire protection, to semiorganized slaves, freemen, or volunteers, and then to the organized public and private fire protection agencies of today that provide protection for all citizens, regardless of where they may live. We know from past chapters that fire is a very real problem within our society; except in its very early stages, it is usually beyond the control of a single individual. All levels of government and private industry have endeavored to control fire; however, the key agency of fire protection, and the one that is most important to you and me, is the local fire department. The local fire department is our first defense against fire and the level at which most fire science students begin their fire protection careers. For these reasons, in this chapter we will explore the purpose and organization of local fire protection.

In addition, we shall review the high points in the history of man's attempts to protect himself and others from fire. History not only provides us with a background of fire protection efforts, it also shows us how we arrived at some of the fire protection methods and traditions of today. Perhaps, we may use history to prevent us from repeating the costly errors of the past.

At the end of this chapter we will have accomplished the following objectives, and you will be able to:

1. Identify some of the historical efforts man has made to protect himself from fire.
2. Identify and define the purpose and scope of fire departments.
3. Identify the basic types of local fire protection agencies and their organization.
4. Identify the main job titles, duties, and basic requirements for the positions within a local fire department organization.

New Terms

Familia Publica
Corps of Vigiles
Paramilitary
Volunteer
Part paid
Fully paid
National Board of Underwriters
Turnout or bunker suits

HISTORICAL FIRE PROTECTION EFFORTS

According to mythology, man was given the gift of fire by Prometheus, who stole it from the gods and was therefore doomed to suffer. As soon as humanity gained access to fire, control over this life-giving yet death-dealing force became a problem. As humans joined together and formed settled communities, fire became a threat not only to any crops they grew, but to the whole group. Evidence of destruction by fire is found in almost all uncovered prehistorical sites and in ancient villages and cities.

The first recorded evidence (that we know of thus far) of organized efforts to protect a community against fire is found in the records of Rome, from circa 300 B.C. At that time a band of slaves, the Familia Publica, kept watch for fire and warned the Roman citizens of this danger. Later, the Familia Publica gave way to what can be considered as the forerunner of modern fire departments.

Under the aegis of Caesar Augustus, the Corps of Vigiles came into being. It was a paramilitary organization (not unlike our fire department) and had ranks and duties for each member. In addition, the men in the fire stations *(castra)* had duties which included watching for fire, warning citizens of fire, and taking positive steps to extinguish any fire. The Corps of Vigiles structure included the following titles, and their modern day counterparts are given for reference:

Roman	*Modern*
Praefectus vigilum	Fire chief
Fire centurion	Assistant chief, battalion chief, or Captain
Quarstionarius	Fire marshal
Siponarius	Engineer
Aquarius	Fire fighter, hoseman, or nozzleman
Nocturne	Alarm operator

It is interesting to note that a nocturne was responsible for sounding the alarm of fire, and the device he used was a trumpet. Later, we will see the trumpet used in another manner in the modern fire department. However, to continue with our overview of fire protection, we move on to the years after the Great Fire of London (1666).

After the Great Fire of London (which burned unchecked for five days and destroyed 373 acres of buildings within the city), the need for organized fire protection was evident. What initially began with organized fire protection groups sponsored or paid by insurance interests led to the establishment of public fire protection (the London Fire Brigade) in 1865. These groups were similar to the volunteer fire department organizations of today as the men were called in the event of fire and had a primary duty of salvaging all they could from a fire. European fire protection organization was then well on its way. Let's turn to similar efforts in America.

Early colonial America was not exempt from the ravages of fire. The colonists brought with them many of the same construction methods and talents that had been used in Europe in general and England in particular. As cities and towns were formed, each individual of the community was expected to respond to the call of fire. Bringing their buckets and swabs, they fought fires in the thatch roofs, tried to prevent fire from spreading to gunpowder stores in the center of villages, and tried to confine fire to one building so as to prevent its spread to adjacent homes across the narrow roads and walkways. Many early American towns required that each citizen respond to fire, and failure to do so could be punished by fine or other penalty.

One of America's first organized attempts to protect citizens from fire involved the night watchmen, who were hired to patrol the streets and, among other duties, warn of fire. Their warning devices consisted of rattles, bells, horns, and other noisemaking instruments. According to historians, in 1711 Boston hired fire wardens not only to patrol but to respond with volunteers to fires. As America grew and its cities expanded, the need for organized fire protection grew.

In 1736, Benjamin Franklin, concerned about fire protection, helped organize one of the first volunteer fire departments, and he is generally accepted as one of the young nation's earliest fire chiefs. In addition, many groups organized themselves into membership societies that fought fires and other volunteer fire fighting organizations. These groups tended to admit only those persons that they wanted, and in many instances it was deemed quite an honor to be invited to join. It also led to rivalry between societies. (Vestiges of this closed membership are found in some volunteer fire departments today.)

Along with early volunteers and membership societies for fire protection, insurance groups, by forming salvage companies, helped cut losses due to fire. Each insurance company had its own symbol, and those homes that were insured were given a fire mark (a metal plate with the company's symbol) to place outside their homes. Then in the event of fire, hopefully, the salvage organization of that insurance company would see the mark and set to work trying to save the contents of the building. Later we will see that the modern fire department now performs salvage as well as other fire protection operations.

Eventually, insurance-sponsored salvage corps and membership societies could no longer meet the total fire protection needs of a community and a professional fire fighting force was organized and began to take over. However, the development of trained and professional fire fighters and departments in America did have its problems, and, as in the past, insurance companies spearheaded the drive to improve fire protection.

Let's look at one instance in the development of current fire departments which occurred after the Great Chicago Fire of October 9, 1871. (The fire is commemorated each year in October during National Fire Prevention Week.) After the fire the National Board of Fire Underwriters made several demands, relating to fire protection, to the Chicago City Council. Among them were demands to reorganize the fire department and remove political influence, to improve apparatus and equipment, and to establish discipline. In addition, a fire marshal's bureau was established to enforce fire regulations. An outgrowth of this confrontation between insurance interests and government was the Municipal Grading Schedule.

Thus, through the efforts of insurance companies and government concern, public fire protection has grown to what it is today; there is still room for improvement and a need for new people in its ranks to keep it moving. At this point, let's move on to the objectives of fire protection and fire departments.

OBJECTIVES OF FIRE PROTECTION

Regardless of the level of government responsible (federal, state, or local) or the size of a public fire protection organization, its goals, objectives, and purpose remain essentially the same. Its purpose: to provide fire protection to all citizens within its jurisdiction. Its objective: and, given the fire departments capabilities, to prevent or hold life and property loss to the minimum. In addition, fire departments provide protection and assistance in emergencies other than fire, such as floods, earthquakes, and other natural disasters. Most fire departments provide at least first aid and/or paramedical assistance to victims of accidents, heart attacks, and in other emergencies where human life is in danger.

Fire departments are also called on to prevent hazardous situations from occurring because of the presence of pesticides, explosives, and toxic and other dangerous materials found throughout our communities. In fact, most citizens will automatically call the fire department for assistance in every crisis except in a situation that obviously calls for police action (e.g., crime, minor accidents).

Thus the services of a fire department are far more varied than just fighting fire and minimizing life and property loss caused by fire alone. Today's fire department provides a many-faceted service and, in essence, its goals are:

- Protecting life and property from fire through fire prevention measures, inspections, and fire suppression.
- Providing emergency care, including paramedical care, for the sick and injured.
- Protecting life and property from loss in hazardous situations, other than fire, which may threaten the community (e.g., flood, earthquake, tornado).

Now that we have a firm grasp on the objectives of a fire department, we will look at the organization needed to accomplish these goals.

ORGANIZATION

This section consists of an overview of fire department organization, since, in later chapters, we will discuss specific components in more detail. Thus, consider this section a guide for future discussion.

As noted in the last chapter, the chief of the fire department is responsible to a local political legislative body (city council, board of directors, fire commission). However, we are now concerned with the inner organization of a department; that is, all those persons and functions under the direction of the fire chief.

The first point to establish is that a fire department, regardless of size, is organized in a paramilitary manner with ranks and a chain of command. Thus, the chief of the department can be likened to a general in the Armed Forces. The chief, like the general, is responsible for the direction, control, actions (good and bad), and safety of all those under his authority down to the newest fire fighter. (In essence, the chief is the top of the pyramid and at the base are the front line troops: the fire fighters.)

Next in command is the assistant chief or deputy chief. When the chief of the department is absent, the assistant chief assumes total responsibility and authority for the fire department. Below this level in the organization, there are two main divisions: fire suppression and fire prevention. (Sometimes administration is considered a third division.)

Fire suppression and fire prevention have essentially the same goals; however, their functions and methods differ somewhat. The fire suppression division has the primary mission of extinguishing fires quickly and efficiently, providing first aid and/or paramedical services, taking care of fire department apparatus and equipment, and alarm and communications systems. On the other hand, the fire prevention division has the primary mission of preventing fires from starting via inspections, code enforcement, and public education, as well as the responsibility for fire investigation and legal action (more on fire prevention in a later chapter). Each of the two divisions is the responsibility of a chief officer (division chief), who is responsible to the assistant chief and the chief of the department.

SUPPRESSION DIVISION

Under these division chiefs are further subdivisions, each charged with specific duties and responsibilities. We will discuss here only the fire suppression division. At each level there are more officer ranks, until we reach the engineer, driver, fire fighter, paramedic, dispatcher level. Some examples of these ranks are:

Chief officers	Division chiefs District chiefs Battalion chiefs
Company officers	Captains Lieutenants
Other ranks	Engineers, Drivers, and Operators Laddermen Nozzlemen Paramedics

(Note that two types of officers are included in the designations above; these designations will be explained in Chapter 6. The ranks in fire prevention, such as inspector, vary with the department.) Each designation or rank in the chain of command has its specific responsibilities and work and refers to an individual responsible to the next highest ranking officer. In addition, there are other designations such as chief's driver or assistant, alarm operator, and mechanic. Obviously, not all fire departments need or utilize all of these various ranks, nor do they necessarily have more than one division; the organization depends on the size and needs of the specific department. We now move on to types of fire departments.

TYPES OF PUBLIC FIRE DEPARTMENTS

There are three basic types of public fire departments in the United States:

Volunteer

Part paid and part volunteer

Fully paid

The sizes of each type of department vary from a one to two man operation up to large departments which have more than 1000 men (e.g., the city of Chicago with 4915 men). Keeping this factor in mind, let's briefly look at the essentials of each type, beginning with a volunteer department.

Volunteer

Throughout the United States, volunteer fire departments have performed and still do perform many fire protection functions. Their responsibilities include small towns, villages, and sometimes urban areas, as well as rural and agricultural fire protection. Some are highly organized and perform many functions, and others provide only minimum fire protection as the need arises. Regardless of their individual size or abilities, their organization remains the same—that is, the chief is in charge.

The chief may or may not be paid, and may receive free housing adjacent to the building containing the department's fire equipment. He is then responsible for receiving and responding to all fire calls within his jurisdiction, as well as the enforcement of fire regulations. The other members of the department may also hold various ranks, and are notified by siren, horn, radio, and/or telephone to respond to a fire call. Most volunteers attend training meetings and drills, and may or may not be paid a small amount of money for each training session and/or fire attended.

Part Paid and Part Volunteer

As can be discerned from the title, a part paid and part volunteer department combines two opposite types of service. This kind of department relies on full-time paid men but is supplemented by volunteers. Usually the paid men are responsible for training, initial response, and upkeep of equipment, as well as related duties, and the volunteers attend drills and respond to calls as needed. In other instances, the department may have a large number of paid men and rely on volunteers as an auxiliary and sometimes political support force. In these organizations, the volunteers respond only to those fires where additional manpower and/or equipment is needed.

The advantage to this type of fire department is that men are on duty at all times and can quickly respond to fires, yet the cost of paying all the men that serve is avoided. Readily available manpower is provided by the volunteers, without excessive cost to the fire protection jurisdiction. In fact, as cities and urban areas face smaller budgets and fire fighting manpower is still needed, this type of department may be used more widely. In some areas public safety departments use policemen to fill in as fire fighters when the need arises and train men to do both jobs, retaining only supervisory personnel in the specialties (e.g., police, fire protection). The chain of command and paramilitary organization remain the same in both instances, but with more full-time men fire protection tasks other than fire fighting alone can be accomplished.

Fully Paid

Fully paid departments are organized and operate much as outlined under the section entitled "Organization." In this type of department all of the men are paid, and the job is their profession, regardless of the type of fire protection involved (e.g., forest, rural, urban). Since the remainder of the book will be primarily about fully paid fire departments, no further discussion is necessary at this time.

From this discussion of different types of fire departments and fire department organization, let's move to determining how to tell who is who in the department, and identifying some of their functions.

HOW TO TELL THE CHIEF FROM THE INDIANS

In addition to being paramilitary in organization, fire departments are usually a uniformed service. All the members of the department wear uniforms and symbols to indicate their rank and/or function. In addition, rank is indicated by the color of apparel (hats, hat bands) and/or the color of metal badges and lapel ornaments.

An example of the combination of symbols and colors is the uniform of a chief of a fire department which will include a white hat with gold metal. For dressy occasions, a white shirt is usually worn in addition to the aforementioned items. The symbols used to indicate his rank are several (usually five) crossed speaking trumpets on both the cap badge and breast badge (Fig. 4.1). In the days before electronic communication, the chief used the speaking trumpet to amplify his orders during firefighting operations; now the speaking trumpet serves as a symbol and a reminder of tradition. In addition to the symbols, the breast badge is usually worked to indicate the name of the department, as well as the wearer's rank (e.g., chief, assistant chief, battalion chief). All chief officer's badges and metal are gold, and the lower the chief's rank the fewer crossed trumpets on his badges.

Fig. 4.1. An example of a chief of department's breast and hat badge. Note the crossed speaking trumpets. Both of these badges are a gold color. Courtesy V.H. Blackinton & Co., Inc.

At the company officer level, the color is silver (badge, cap badge, and hat band and/or chin strap) and the speaking trumpets are not crossed (Fig. 4.2). Depending on the department, the captain's hat may be white or very dark blue; the lieutenant's caps are usually dark blue. The number of trumpets indicates rank, with a captain having two parallel trumpets and a lieutenant having a single trumpet. Again, wording on the badge describes the exact title and department. Variations in these company officer symbols are used where truck companies have their own officers; axes are used as symbols in the same manner as the speaking trumpets.

Fig. 4.2. An example of a captain's breast and hat badge. Note the two speaking trumpets. Both badges are a silver color. (Courtesy V.H. Blackinton & Co., Inc.)

At all other levels the badges remain the same, with hydrants, pikepoles, ladders, and nozzles used as symbols, and the title, such as engineer, inspector, and fireman, indicated. The color of the badges is usually silver, but the cap bands and/or chin straps are black.

Other symbols are used to indicate the number of years of service (e.g., one small star on the sleeve for each year of service) as well as any special skill (e.g., first aid instructor). Fire departments have traditionally dressed in very dark navy blue, but the specifications of the style of uniform and other clothing are dictated by the department's regulations.

The use of colors on the fire ground should be mentioned. The fire scene is no place for dress uniforms; therefore, turnout, or bunker, clothing is used to protect the wearer from heat and water. The helmets, coats, and, in some departments, the pants, are colored to indicate rank. Again, white is the traditional color for chief officers' clothing, including both the helmet and coat. Wording may be added to indicate rank and last name.

Captains and other company officers may or may not have turnouts of a different color to indicate rank, but the helmet is usually marked so they can be picked out on the fire ground. The fire fighter's turnouts are usually unmarked, except for reflective tape and perhaps the last name; the colors used depend on the department, but all the turnouts below officer ranks are the same color.

The use of colors and symbols both at the fire and when on duty serves two basic purposes, which are to continue tradition and to mark the individual so that he can be readily picked out to transmit orders between his superior officer and the men below him. In addition, as with any uniformed service, colors and symbols give the wearer a sense of individuality and accomplishment as he advances in the ranks and becomes eligible to wear marks that indicate his achievement.

Before we leave the subject of uniforms, it should be noted that there has been a movement to use civilian business clothing for members of the fire prevention bureau and arson squad (if any), and for the chief of the department, who, for various reasons, deal with the public on a continual basis. If not in uniform, these people carry their badges and identification on their persons, but they are not displayed. The motivation behind the move to regular business clothing for these individuals is a feeling that you can deal with the public more effectively without outward symbols of authority—individuals rather than badges are effective. However, not all fire departments believe in this philosophy, and it is not in widespread practice at this time. Tradition plays a large role in the fire service.

SUMMARY

As man began to live in primitive groups, fire became a greater danger to himself and others of his kind. However, before Rome we know of no sustained organized efforts to protect man from raging fire.

The first recorded organized fire protection discovered thus far was the use of slaves for the Familia Publica. Later, the Romans penchant for organization and protection of its citizens gave rise to what was possibly the forerunner of the modern fire department, the Corps of Vigiles. The Corps of Vigiles was a paramilitary organization that assigned specific ranks and duties, including the equivalent of the modern fire Chief, fire marshal, engineer, and fire fighter. They also had firehouses *(castra)* and an early warning system manned by the nocturnes.

The next major step in organized public fire protection occurred in Europe after 1666. Insurance interests organized salvage groups after the Great Fire of London, and these groups were the forerunners of the modern London Fire Brigade.

In America, the lessons about fire were relearned in the early colonies, where fire protection was an individual responsibility. Each citizen was responsible for fire fighting and had to respond to any fire with his bucket or swab in hand. Combustible construction, gunpowder storage in the center of towns, and narrow streets made conflagrations a rule rather than an exception. As the towns grew into cities and the cities hired watchmen to patrol the streets to warn of danger, the first steps that led to the eventual organization of public fire protection began.

By the 1700s groups of men had banded together either as part of a membership group or as volunteers to help provide fire protection. In 1736, Ben Franklin became one of the young nation's first fire chiefs, and he helped promote organized fire protection.

To the membership fire fighting societies and organized volunteers, insurance interests added salvage companies to protect their insured clients' property. Fire insurance marks were used to identify homes that carried insurance; when a fire occurred, the various salvage companies would salvage that property that was insured by their company. Eventually, salvage companies and membership societies gave way to organized public fire departments (volunteer and paid) that provided needed fire protection. A further push to better organize fire protection came after the Great Chicago Fire in 1871. Several insurance companies banded together under the name of the National Board of Fire Underwriters, and this group demanded that fire suppression and prevention be changed; out of the Chicago confrontation came the fledgling Municipal Grading Schedule.

The modern fire department is a structured entity that strives within its capabilities, to reduce loss of life and property by fire to the minimum, via prevention and suppression efforts.

In addition, first aid and paramedical assistance is expected from and provided by the modern fire departments; their efforts include aid during natural and man-made disasters. To accomplish these roles, the fire department must be well trained and organized.

Traditionally, a public fire department is organized in a paramilitary manner with a chain of command, ranks, and specific duties assigned to each rank. The chief of the fire department is the top ranking officer and is responsible for the overall action and efficiency of its members. Under him in the chain of command are various chief officers and company officers. Orders are passed down through the ranks and requests are channeled back up again through the next highest ranking officer. All members of the department must follow the chain of command, from the rookie fire fighter to the chief of the department.

The same structure is used in the three basic types of public fire departments: volunteer, part paid and part volunteer, and fully paid. The essential difference between the three types lies in the kind of manpower available, the number of persons on duty at all times, and what individuals receive as compensation for duties performed.

Among the uniformed fire departments, colors and symbols are used to denote rank and title. Gold and white are usually reserved for the chief officers, silver for the company officers, and dark blue for other ranks. In addition, the speaking trumpet is used on badges to further denote rank, and other symbols such as ladders, nozzles, and hydrants are used on badges for ranks below the officer level. Colors and symbols follow tradition and also serve as identification for both the wearer and officers in charge at the fire scene. In certain areas of fire protection (i.e., prevention and investigation) in some departments, uniforms are replaced by normal business clothing; yet, ranks are still used, even though they are not constantly displayed.

REVIEW QUESTIONS

1. What were the roots of modern fire protection?
2. How was fire protection provided in early American cities?
3. How have insurance interests affected fire protection?

4. What are the overall objectives of modern fire protection?
5. What is meant by the term "paramilitary."
6. List the fire department positions in order of rank and note the particular symbols and colors used for each rank.
7. There are several types of fire department organizations; list them and describe the differences between the types listed.
8. What is meant by the term "chain of command"?

REFERENCES

National Fire Protection Association, *Fire Protection Handbook,* 14th ed., Boston, MA, 1976.

William K. Bare, *Fundamentals of Fire Prevention,* John Wiley & Sons, New York, NY, 1977.

PART II

II FIRE PROTECTION SYSTEM COMPONENTS

chapter 5

5 APPARATUS AND EQUIPMENT

Any system, regardless of type, consists of components that make the system function (either effectively or ineffectively). In a fire protection system (in this instance, a fire department) there are many subsystems and many components; one of the essential components is composed of the apparatus and equipment. Without modern apparatus and equipment, the ability of a fire department to cope with emergencies would be seriously hampered, if not rendered almost ineffective.

In this chapter we will define and discuss the major types of apparatus used in a modern fire department: basically, rolling stock (e.g., pumpers, ladder trucks, tankers). We will review some of their history, and identify some special apparatus, such as that used in airports.

Under the heading of equipment we will discuss those components that are carried on the apparatus, including items such as the personal equipment of a fire fighter, used at a fire scene (turnouts or bunker suits). The discussion will be limited to those items that are common to most fire departments and those that a new fire fighter would have to be familiar with in order to function at a fire scene.

At the end of this chapter you will be able to:

1. Identify specific types of apparatus and their differences.
2. Identify the main functions of each type of apparatus discussed.
3. Identify specific types of common fire department equipment and their use.
4. Identify the kinds of personal equipment used for fire fighting and their function.

New Terms

Apparatus	Forcible entry tools
Equipment	Salvage covers
Pumper	Smoke ejector
Tanker	Turnout or bunker clothing
Engine	Elevating platform (snorkel)
Truck	Crash truck
Aerial	Booster/chemical hose
Female coupling	Hard suction hose
Male coupling	Soft suction hose
Double female	Combination nozzle
Double male	Master stream appliances
Wye	Ground ladder
Siamese	Extension ladder
Reducer	Roof ladder

APPARATUS

Apparatus, within the context of a fire department, means those vehicles that are used for fire protection and transportation. Commonly, the public refers to these vehicles as fire engines and fire trucks, regardless of the vehicle's function. As we will see, these vehicles are much more than trucks; each type and design has specific purposes and functions in addition to transporting fire fighters to the scene of an emergency.

Let's start our discussion with one of the most common pieces of apparatus—and the one with the longest history—the pumper.

Pumper

The first pumpers were developed in England in the 1700s. They relied on men to move them to the fire scene and utilize the equipment (leather buckets). As the pumper was developed further, it became capable of taking water via suction from a cistern or other supply and increasing the pressure to provide a solid stream of water from a nozzle mounted on the pumper itself. Again, man power was needed to make the pump work, and as the energy of the men manning the pump handles lessened, so did the effectiveness of the hose stream. Furthermore, until the development of the first fire hose (leather) in England, the pumper had to be placed close to the fire to allow a water stream from the affixed nozzle to be effective, which made the work very hot for the early fire fighters.

As the technology of the industrial revolution grew, fire fighting apparatus improved, and horses and machines took over locomotion and pumping. In the 1870s, the more progressive fire departments were using steam, not only to pump but to move the pumpers to the scene of the fire; gradually the horse was replaced, but not displaced from the history of the Fire Service.

By the early 1900s, automotive fire apparatus came into prominence, and in many areas the horse-pulled apparatus was raced against the horseless pumpers for the last time. Over the years the pumper has been refined. In the future, it may be automated in many respects. However, let's examine the pumper as it is today.

The modern pumper (Fig. 5.1) is a machine that performs many functions other than pumping water from a hydrant to a nozzle. The pumper has become a combination of a pump, water carrier, hose carrier, and an equipment carrier of large and small tools. (In the not too distant past, a hose wagon carried the hose, and the pumper was a separate piece of equipment.) Although pumpers vary in design and use, the "average" pumper today is a triple combination piece of apparatus. Triple combination means that it carries a supply of water in a tank, in addition to a pump and hose. To expand a bit further, a pumper is a gasoline or diesel-driven piece of automotive apparatus which carries a pump rated to deliver a specific number of gallons per minute

Fig. 5.1. An example of a pumper/engine (triple combination). The ladders are carried on the opposite side of the engine, out of view in this photo. (Courtesy Crown Coach Corp., Los Angeles, CA.)

(1000 gpm, for example). It carries a limited number of ladders (usually a 24-foot extension, 14-foot roof, and a 10-foot collapsible or attic ladder), hose (1500 feet of 2½–inch diameter, 400 feet of 1½–inch diameter, and smaller chemical or booster hose of ¾–inch diameter), and a water tank (with a minimum capacity of 300 g). Other equipment includes portable fire extinguishers, tools, nozzles, and breathing equipment, which is stored in compartments on the pumper chassis.

Of course, apparatus and equipment can and do vary from department to department. The capacity of the pump may vary from a minimum of 750 to 2000 gpm. Between the two extremes pumps are sized at 250-gpm increments. The amount of hose mentioned above is the minimum amount recommended in the National Fire Protection Standard No. 19, "Automotive Fire Apparatus," which serves as the nationally recognized standard for most fire apparatus. (Aircraft Rescue and Firefighting vehicles—NFPA No. 414)

There are other adaptations of the basic pumpers which are

designed for special use, such as brush and wildland fire fighting apparatus, airport crash apparatus, and rural fire protection equipment; for these vehicles, among other necessities, a larger water-carrying capacity is required. For pumpers that operate in "off the road" conditions, such as those used in wildland fire fighting, four-wheel (all-wheel) drive is utilized. Another variation on the basic pumper that is gaining acceptance is an adaptation of the old water tower. Pumpers are now being constructed with a telescoping water tower that provides elevated hose streams directly from the pumper at heights of from 55 to 75 feet, depending on the design capabilities. This tower, such as the one called "Tele-Squirt," may also have a ladder attached, so that it can serve as a small version of an aerial truck. In essence then, pumpers can be designed to an individual fire department's needs and specifications; however, they still must meet the minimum safe construction standards and equipment requirements (more details are presented later in the section on apparatus construction). A piece of apparatus that works in conjunction with a pumper is the tanker.

Tanker

Tankers are vehicles equipped to carry large volumes of water so that they can act as mobile water supply in areas where readily available fire fighting water is scarce (e.g., rural areas). The water-carrying capacity of a tanker varies; the minimum capacity is 1000 gallons. They can be built to carry up to 5000 gallons or more, but this requires heavy construction to hold the weight of the water, and the larger tankers are usually tractor-trailer assemblies. Tankers can and do carry other fire fighting equipment, but for the most part their main function remains that of supplying water to pumpers.

Another variation of the straight tanker is a combination achieved by adding a pump, or a combination pumper-tanker. In this instance, the apparatus is designed to function both as a water supply and as a pumper, and carries hose and other fire fighting equipment.

We now move from our discussion of pumpers and tankers to the aerial ladder trucks and elevated platforms.

Aerial (Aerial Ladder Truck)

As buildings became taller, the need for ladders longer than those that could be lifted by man power alone (50 feet or shorter) became evident. Over 100 years of use and refinement has led to the aerial ladder apparatus of today (Fig. 5.2), and new needs are changing the current design. The first aerial ladders had to be manually cranked up and down; later, springs and air hoists moved the heavy wooden ladders.

The modern aerial ladder is no longer made of wood, nor is it operated by heavy springs. The wooden ladder has been replaced with materials such as steel and aluminum, and an aerial operator can move the ladder with a slight pressure of his hand. The vehicle's engine along with hydraulic lifts and cables, lifts and extends the three to four sections of the ladder.

The common lengths of today's aerial ladders, when fully extended, are 85 and 100 feet. In the past, 65- and 75-foot aerial ladders were common on new aerial trucks, but are now rarely produced. The demand is for longer lengths, due to the greater heights found in new construction. Although it is possible to find 150-foot long aerial ladders, they are not common, and the production of even greater lengths is probably not possible with our present technology. Due to safety requirements and the limits to the size of the base (the truck chassis itself, with the addition of manually or hydraulically lowered outriggers), methods other than increasing the length of the aerial ladder for access to upper stories of buildings must be relied on.

Aerial ladders can be mounted on either a single one-piece chassis or on a tractor-trailer assembly. The length of the aerial and amount of equipment carried, plus the desires of the fire department, are the main factors determining the chassis design. For those aerials that have a tractor-trailer assembly, two sets of steerable wheels are usually provided: one set of steering wheels at the front of the tractor and one set at the rear of the trailer, guided by the tillerman. Thus, when responding to a fire the driver-engineer is one part of a two-man team; his partner is the tillerman, hard at work in the rear. Because it is a team, training and communication is important. The tillerman is provided with some way of communicating with the driver (bell, buzzer, inter-

Fig. 5.2. An example of an aerial or aerial truck. This one is a 100-foot aerial ladder shown fully extended from its rear mount. (Courtesy Crown Coach Corp., Los Angeles, CA.)

com) to indicate what has to be done (e.g., back up, ready, all clear). Like the driver, the tillerman is provided with a windshield, a seat with seatbelt, and a steering wheel to help control this long piece of apparatus (Fig. 5.3).

In addition to the aerial ladder, an aerial ladder truck also carries an assortment of other ladders (ground ladders) and equipment. The ground ladders (those operated by man power) make a total, in aggregate length, of a minimum of 163 feet and may add up to 400 feet. An example of ground ladders totaling

Fig. 5.3. A view of the back of a tractor-trailer aerial ladder truck. Note the tillerman's seat and the number and location of the ground ladders. (Courtesy Crown Coach Corp., Los Angeles, CA.)

163 feet, the minimum length that meets the NFPA standard, is usually as follows:

- Four extension ladders, one each of 40, 35, 28, and 14 feet.
- Two single ladders with roof hooks, or two roof ladders, one each of 20 and 16 feet.
- One collapsible ladder or attic ladder 10 feet long.

Figure 5.3 shows how these ladders are usually carried, and how they are accessible from the rear of the truck.

In addition to ladders, an aerial carries pikepoles, salvage equipment, lighting equipment, breathing equipment, ropes, and nozzles for the elevated hose stream equipment. The elevated hose stream equipment is detachable and must be set up at the fire scene before it can be put into operation. This means that the nozzle must be attached to the aerial ladder before it is raised, the hose must be laid out to extend from the ground to the nozzle, and someone must climb the ladder to operate the nozzle once a water supply and pump is attached (usually provided by a pumper).

The aerial ladder truck serves several functions that would be difficult or impossible to carry out by pumper alone. Some of these functions are:

- To assist in placing men and equipment on the roof or upper floors of a building so that they can carry out tasks such as ventilating smoke and toxic gases from a building fire.
- To aid in the evacuation and rescue of occupants from upper stories of buildings and other elevated structures beyond the reach of ground ladders. (They are also used to rescue people who have fallen over cliffs.)
- To provide large hose streams from elevated heights via the ladder pipe (the detachable nozzle and hose assembly carried on the truck).
- To carry needed equipment that cannot be carried on pumpers and other apparatus, such as additional self-contained breathing equipment, portable pumps, electrical generators, salvage equipment, and assorted ground ladders.

Another variation of the standard aerial configuration is accomplished by adding a pump so it may supply its own water for elevated hose streams and other fire fighting operations. The next piece of apparatus discussed is similar in function to the aerial but of a different design: the elevating platform, or snorkel.

Elevating Platform (Snorkel)

The elevating platform truck differs from an aerial in that rather than having an extendible and retractable ladder, it has a platform on the end of a telescoping or articulated boom (Fig. 5.4). The most common lengths for elevating platforms are 50 to 90 feet, measured from the ground to the platform when fully extended.

The main differences between an elevating platform and an aerial ladder, other than structural differences, are as follows:

Elevating Platform	*Aerial Ladder*
Can be controlled from either the platform or from the base of the boom.	Can be controlled only at the base of the ladder.
Contains built-in piping and nozzles for elevated hose streams.	A ladder pipe must be attached when needed, in addition to stringing the hose up the ladder.
The platform forms a stable base for men and equipment and occupants of burning buildings when they are rescued.	Requires that all equipment be moved by people climbing up and down the ladder.
Requires more clear area for operation due to the arc of the boom.	Can be used in confined areas, and has a greater variety of lengths of extension (e.g., 100 feet, 150 feet).

As with any type of specialized equipment, some items perform better in certain circumstances than others, and this is true of the aerial versus the elevating platform. Some departments use both types of apparatus; however, both are very expensive, and most medium-size departments have to choose one or the other for their needs. As with the aerial, an elevating platform can be pro-

Fig. 5.4. An example of an elevated platform ("snorkel") truck. Note the differences and similarities between this and the aerial. (Courtesy Crown Coach Corp., Los Angeles, CA.)

vided with a pump and other equipment to form a multipurpose piece of apparatus. Elevating platforms carry the same types of ladders as the aerial, and in the same basic configuration.

Special Apparatus

In addition to the main types of apparatus thus far discussed, a fire department may have other rolling stock for special situations. Included in this group of apparatus are rescue and paramedical units, salvage and light units, fireboats, airport crash trucks, and smaller vehicles such as chiefs' cars. Let's examine each of these briefly and note their functions, beginning with the rescue and paramedical vehicles.

Rescue and paramedical units range from standard ambulances to a specially equipped small (1-, 3/4-, and 1/2-ton) trucks. These units are designed to get to the scene of an emergency quickly and provide the needed equipment to care for the victims of fires, auto accidents, drownings, heart attacks, and any other catastrophes requiring immediate first aid or paramedical assistance. These vehicles may also be equipped with fire extinguishing equipment, electronic treatment aids, forcible entry tools, and other small tools and equipment which might be needed by their crews. Some of them are designed to transport victims, but for the most part, rescue vehicles are concerned not with transportation of the injured but with providing emergency aid; the transporting is accomplished by other vehicles, such as private ambulances.

Salvage and light units are like rescue vehicles usually small. These vehicles are designed to carry special equipment (salvage covers, water vacuums) or, if designed for lighting units, they carry electrical generators, large and small flood lights, and other equipment that is necessary to provide electrical power and illumination at the scene of a fire or other emergency.

Fireboats are used where waterfront properties and waterborne vessels must be protected. The size of fireboats varies greatly, from a small 16-foot open boat with a small pump and one or two nozzles to a large tugboat-size fireboat carrying a full crew of trained seamen and fire fighters, and capable of pumping 10,000 gallons per minute and supplying a large assortment of nozzles.

Airport fire protection apparatus (crash trucks) are essentially a specialization of the basic pumper we described earlier. However, because of the kinds of fuels and materials encountered (e.g., jet fuels, combustible metals that react with water), special extinguishing agents are required. Thus, rather than tanks of water, foam and chemical extinguishing agents (e.g., carbon dioxide, dry chemical) are carried.

Because of the intensity of aircraft fires and the lack of readily available extinguishing agents from accessible sources, crash trucks must be capable of providing effective fire fighting streams for at least five minutes. Therefore, as would be expected, a crash truck must be big and heavily constructed (eight tons or more), to carry the weight of the large quantities of the essential materials. Smaller vehicles are used for support functions, such as carrying rescue equipment and serving as command vehicles which coordinate the larger crash trucks' attack. Tanker type crash trucks are also used as "nurse" vehicles when needed to supply additional extinguishing agents.

Frequently, the heavy airport apparatus is modified even further because of the terrain it must cover; it must be capable of arriving rapidly at the crash site, regardless of weather or ground conditions (mud, snow, slippery jet fuel). A crash truck has a high ground clearance and positive drive to all wheels which increases its maneuverability. Most crash truck designs remind me of armored vehicles; they have a boxy appearance and movable turret nozzles that protect the vehicle, as well as provide fire fighting extinguishing streams. They also have hand-held nozzles which enable personnel more easily to attack a fire at several points.

Other specialized pieces of fire fighting apparatus that should be mentioned are the aircraft employed in wildland fire fighting, and the helicopters used both for fire fighting and command. Both large multiengine and single-engine aircraft are used to carry fire-retardant chemicals, which are dropped from the air to contain wildland fires. Many of these "bombers" are vintage propeller-driven bombers from the past that have been adapted for fire fighting use. Smaller fixed-wing aircraft are used to coordinate the larger bombers at a fire, and serve as spotters and equipment carriers in wildland fires. Helicopters are used both in municipal and wildland fire protection for transportation, spotting, and dropping fire retardants and water. Very few municipal fire departments can afford or need aircraft, but the larger ones, such as those in southern California, use aircraft for municipal fire protection. However, only the wildland fire protection services (e.g., National Forest, State Forestry) use aircraft to any significant extent.

From our discussion of aircraft, crash trucks, pumpers, and aerials we move to the standard-size cars, station wagons, and pickup trucks that fire deparments use.

The cars are usually reserved for use by the chief, other command officers, and fire prevention officers and fire investigators. They serve as mobile command posts, personnel transportation, and as carriers of assorted small tools and other fire fighting and rescue items. The small pickups (1/2 and 3/4 ton) may be provided with small gasoline-powered portable pumps and hose for fire patrols, or can be used to transport other special equipment, such as special trailers and dirty hose, to and from the fire scene.

Trailers have been adapted by some departments to carry items such as compressed air which is used to refill self-contained breathing equipment tanks; in addition, they can serve as mobile lighting equipment carriers. Small departments find trailers more economical than a piece of motorized equipment which requires a driver.

All manned fire department emergency apparatus must be provided with warning lights, radios for two-way communication, and warning devices such as sirens, air horns, or bells.

Now that we are familiar with the different kinds of apparatus and their basic functions, let's take a closer look at the construction of pumpers and aerials and examine the testing and the criteria for acceptance of these vehicles.

APPARATUS CONSTRUCTION AND TESTING REQUIREMENTS

We are now going to build a piece of apparatus in words and with the help of pictures to give you an idea of what is behind all the chrome and bright paint. This should help you to identify the basic components hidden within.

We will build a "typical" pumper from the ground up, beginning with the chassis (Fig. 5.5). The foundation for any type of apparatus is the chassis; the remainder of the vehicle is added to it. Depending upon the specifications of the fire department, the chassis can either be a standard chassis (e.g., Ford) or a special chassis designed by the fire apparatus manufacturer.

The first additions are the power plant (Fig. 5.6), drive train, and all the supporting systems connected with the engine and transmission (e.g., cooling system, fuel system). The power plant can use either gasoline or diesel-fuel; diesel is gaining favor because it is cheaper, and diesel motors have the ability to run efficiently for long periods, as is required in pumping operations.

Next, the main pump is added (Fig. 5.7) and connected to the transmission and booster pump. In our illustration the pump is mounted in the middle (amidship or center mounted). The pump can also be mounted in the front of the apparatus (front mounted). After the pump is mounted, the plumbing for the pump and hose inlets and outlets is made up (in a manner similar to plumbing in a house), and the pipes are secured in position on the frame. The plumbing is determined by the apparatus specifi-

Fig. 5.5. The chassis of a pumper.

cations, the design of the body, and where the hose connections to the pumper are desired. The main water tank (Fig. 5.8) is connected to the pump (Fig. 5.9*a*). Note in Fig. 5.9*b* the cross-pieces (baffles) in the water tank, which prevent the water from sloshing so much that its weight moving back and forth and side to side would seriously hamper steering and the safety of the apparatus while it is being driven.

Next, the metal for the cab is added (Fig. 5.10); then the side cabinets, hose bed, bumpers, tailboard, gauges, and all the other features are added, painted, and chromed, until we have the finished product (Fig. 5.11). However, before it is delivered to a fire department and accepted, it must be tested by the manufacturer to see if it meets specifications relating to pumping, engine operation, and other functions.

Fig. 5.6. The engine and drive train are added to the chassis.

Fig. 5.7. The main pump is added.

Fig. 5.8. The main water tank and the plumbing for all the pump outlets and inlets are now in place.

Testing

A new piece of apparatus must be tested and meet all specifications before it is ever put into a fire station to respond to fires. The apparatus has to meet federal specifications for braking, construction, and other safety features set by the Department of Transportation. It must also meet the Federal Environmental Protection Agency's standards for pollution emissions from the motor.

Futhermore, standards must meet those in the specifications given by the fire department; which in turn are based on the apparatus specifications listed in the National Fire Protection Association's Standard No. 19, "Automotive Fire Apparatus." This document provides detailed requirements for all types of apparatus, as well as minimum amounts and kinds of equipment that should be carried on each piece of apparatus.

Other standards are provided by the nationally recognized testing agencies (such as Underwriters Laboratories); these include data for certification tests of pumps. The pump certification

Fig. 5.9. *(a)* A closer look at the pump plumbing and the main tank.

test, which determines pumping capacity, is one of a series of certification tests performed at the factory. Other tests include:

1. Road test of acceleration, braking, and maneuverability.
2. Strength tests of aerial ladders and elevating platforms.
3. Stability testing for aerials and elevating platforms during operation.

Upon delivery of the apparatus to the fire department, the same tests (the acceptance test) are usually conducted (e.g., pumping test), and copies of prior certification tests are handed over. The acceptance tests are usually witnessed by the chief of the fire department, or his representative, and other local government officials; if all goes well, the apparatus is accepted, and then the process of getting it equipped for fire fighting duty is begun. We now move to our next topic of discussion: fire department equipment.

Fig. 5.9. *(b)* The view from the front of the apparatus. Note the baffle plates in the main tank.

EQUIPMENT

Our discussion of equipment in this chapter will be limited to that found on apparatus and personal protection equipment used during a fire or other emergency. Furthermore, although a wide and varied assortment of equipment may be used by any one fire department, we will discuss only those items that are commonly

Fig. 5.10. The cab is added.

carried on all types of apparatus and should be readily identified by a new fire fighter. Let's begin by examining hose and accessories.

Hose and Accessories

As noted in the section on pumpers, hose is carried not only on pumpers but on most combination apparatus. The hose carried today varies considerably from the first hand-sewn leather hose available to early fire departments. Leather has now been re-

Fig. 5.11. The finished product, complete with chrome and paint. (Courtesy Crown Coach Corp., Los Angeles, CA.)

placed by rubber, cotton, linen, and synthetic materials. However, the function of hose continues to be the transporting of water or other extinguishing agents from one location ot another.

Modern hose is manufactured quite differently than its leather counterpart. Today it is made in different diameters, lengths, and materials for various uses in fire fighting; for example, hose is used for standpipes in buildings. Other examples of different types of hose and their construction are of interest to us at this time.

One common kind of hose, constructed of rubber with a reinforcement of fabric (commonly called booster or chemical hose), is made with either a 3/4- or a 1-inch diameter, and is carried on reels. This hose is used to get water to small fires quickly by using the booster pump on the apparatus to take water from the water tank. Most pumpers carry at least one or two reels of booster hose, and each reel carries from 200 to 300 feet of hose.

Another type of hose (commonly called "fire hose") has a rubber interior sleeve and a single or double jacket of cotton or polyester fabric on the exterior. This hose, carried in large quantities in the hose bed, has diameters of 2½ or 3 inches or larger and is used to supply water from a hydrant or other source to the pumper, and from the pumper to smaller diameter hoses (1½, 2½ inch) and then to hand lines, or directly from the hydrant to the hand lines. The most common lengths for both the larger diameter hose and the smaller hand lines are 50 and 100 feet. There is a male coupling on one end and a female coupling on the other (Fig. 5.12). These couplings allow hoses to be connected to one another or to other hose outlets and inlets on the pumper, hydrants, and private fire protection systems (e.g., automatic sprinkler systems). Without couplings it would be difficult to attach other hose fittings, such as nozzles.

Another type of hose, the suction hose (hard and soft), is of large diameter (4 to 6 inches) and is used to draft water from cisterns, ponds, lakes, and tanks. The "steamer" hydrant has a special large-diameter outlet for attaching large-diameter suction hose which permits a greater water supply to be delivered to the pumper.

Hard suction hose is semirigid, and has a rubber inner liner with wire reinforcement, and layered rubber and fabric on the exterior. It is usually carried in 10-foot lengths on the side of the pumper (see Fig. 5.1); and, when needed, it is connected to the large-diameter pump inlet (Fig. 5.13). If a lake or other natural body of water is used for the supply, a strainer is attached to the end of the suction hose to prevent debris and rocks from entering the pump. When the suction hose is used in this manner, the pump actually creates a vacuum to pull the water up the hose; therefore, suction hose connections must be tight fitting and cannot have any points of air leakage.

Soft suction hose is in construction basically similar to other

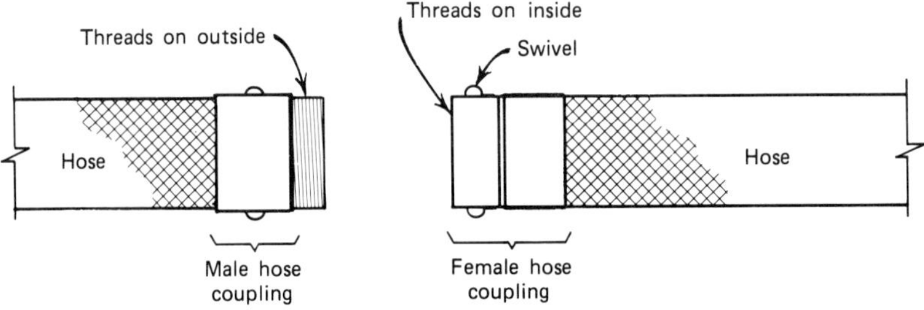

Fig. 5.12. The opposite ends of a length of hose showing the male and female couplings.

Fig. 5.13. A view of the pump on an engine/pumper showing the hard suction hose, part of the booster hose, and the pump suction inlet. (Courtesy Crown Coach Corp., Los Angeles, CA.)

rubber-lined, jacketed hose. Soft suction hose is highly flexible, and is used primarily as a connection between a pumper and a hydrant or similar water supply.

Unlined hose (hose without an inner lining of rubber) is used for special applications where weight is a critical factor and/or cost is important. Two examples of common uses for unlined hose are (1) for fighting wildland fires where fire fighters must carry hose by hand to remote areas and weight is critical; and (2) in private fire protection systems which depend on wet standpipes and yard hydrants. Since this hose does not have a rubber inner liner it is subject to leakage until the fabric (cotton, linen) swells from the water passing through and the minute holes in the fabric weave are closed off. This leakage is one disadvantage of unlined hose; another is greater friction. The friction is greater due to the rough interior of the hose which causes turbulence in the water, which in turn obstructs and slows the passage of the water from the source to the nozzle. The smooth rubber surface of lined hose reduces turbulence, and therefore lessens the friction. To achieve further reduction in friction, greater diameters of hose plus water additives are utilized; fire departments are using larger-diameter (3-, 4-, and 5-inch) hose for supply lines.

Hose Couplings, Nozzles, and Other Fittings

If all fires were the same in all respects, the need for various hose connections and other equipment would be reduced; however, that is not the case. As we have discussed, hose is made in different diameters and lengths, with couplings provided so that hoses can be connected, if necessary, to give greater lengths of hose. Hose may be coupled to other items as when it serves as a line between a pumper and an automatic sprinkler system, between a pumper and another pumper, or to transfer water from a source through a nozzle for an effective fire stream. Consequently, the hose couplings are very important. If a coupling is damaged or missing the hose will not be of much value at a fire.

Until recently, most hose couplings were made of brass, and they added greater weight to the hose. Furthermore the couplings may be damaged by dropping, which can foul the exposed threads of the male end or distort the round swivel of the female end so that it sticks; in both instances, the damage will make the connection of the hose difficult if not impossible. To aid in reducing some of these difficulties, hose couplings have been manufactured of metal alloys that are stronger than brass so that they have a greater resistance to physical damage. These alloys are also lighter in weight, which makes the fire fighter's job of moving hose a bit easier. Still, the threads must be protected, and the couplings checked frequently to ascertain the condition of gaskets and insure smooth operation of the female coupling. Well-maintained couplings will insure connection of hoses with maximum speed and minimum water loss. Since the standardization of hose thread (National Standard), the need for adapters of one type of hose thread to another has been reduced, although adapters may still be carried on the apparatus. However, other hose fittings are carried on the apparatus that deserve our attention.

Hose fittings are made of the same materials as hose couplings and are designed to suit different needs. The following are some examples of the common kinds of fittings carried on apparatus (Fig. 5.14) and their primary uses:

Fig. 5.14. Examples of hose fittings carried on fire department apparatus. Note the arrows on the siamese and the wye which indicate the direction of water flow. (Courtesy Akron Brass Co.)

Name	*Use*
Double Male	To connect two female couplings together; it has exposed hose threads on either side.
Double female	To connect two male couplings together; it has two swivels with interior threads on both sides.
Reducer	To reduce, by reducing the diameter of the thread, so that a smaller diameter hose line can be connected to one of a larger diameter or to a larger diameter outlet.
Adapter	To allow connection of standard hose thread couplings or fittings, by "adapting" a hose thread on an outlet or hose.
Siamese	To join two hose lines to form one hose line.
Wye	To split one hose line into two.

As is evident from the examples, the names of the various fittings give you clues as to their uses. Fire departments carry several types and sizes of these and other fittings in cabinets on the apparatus to give them maximum flexibility in fire fighting operations. In addition to the fittings, an assortment of nozzles are required.

Nozzles

The number and type of nozzles (Fig. 5.15) carried on the apparatus varies, but almost all departments use the following basic kinds:

1. Straight stream nozzles (with and without shutoffs).
2. Combination fog (water spray) and straight stream nozzles with shutoffs.
3. Large-diameter (orifice/tip size) master stream nozzles (both straight stream and combination).

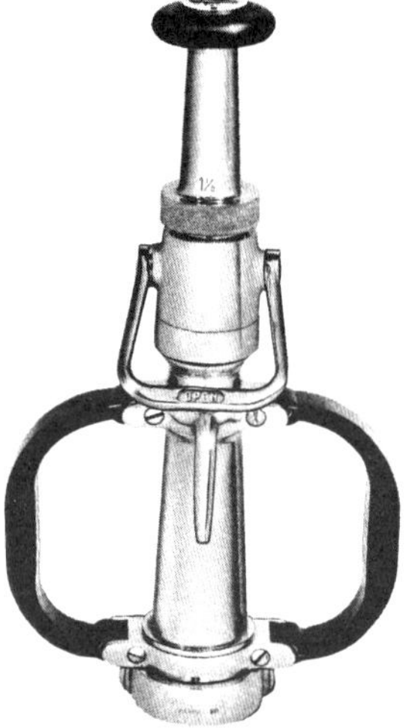

Straight stream

Combination fog and straight stream

Fig. 5.15. Examples of hand held nozzles used by fire departments. (Courtesy Akron Brass Co.)

Departments may also have some special nozzles, such as foam applicators with eductors and/or proportioners (which allow the mixing of foam with water for flammable liquid spills) and cellar nozzles. Each type of nozzle performs certain functions and, again, for the purpose of flexibility, several different types are carried on the apparatus. Lets's examine the basic kinds of nozzles in the order listed.

The straight stream nozzle performs just as the name implies—it throws a straight solid stream of water from the tip. The orifices of nozzle tips for hand held hose lines are available in many sizes but are usually 1, 1⅛, 1¼, and 1⅜ inches in diameter. They are interchangeable and an assortment is carried on the apparatus.

The advantage of a straight stream nozzle is in the length of water stream and reduced friction. However, straight stream nozzles usually require more manpower to handle them and can, if used improperly, create extensive water damage.

Combination fog and straight stream nozzles have the advantage of adaptability, and allow for a choice of the type of stream used. Water spray or fog has the advantage of rapidly absorbing heat (which is a factor in extinguishing a fire) and inflicting less water damage because the major part of the water used is turned to steam. In addition, a spray pattern requires less manpower because there is less nozzle reaction. (Nozzle reaction is the force pushing on the fire fighter while he holds the nozzle. The force used to push water out of the nozzle creates an equal and opposite force which pushes the nozzle backward.) Where required, a combination nozzle can be set for straight stream by the fire fighter holding the nozzle. Some combination nozzles also allow the fire fighter to change the spray pattern and adjust the number of gallons per minute flowing from the nozzle. Fog and combination fog and straight stream nozzles are made in various sizes, both in orifice diameter and at the hose connection to fit different hose diameters (e.g., 1½– and 2½–inch hoses), and hose stream needs.

Master stream nozzles are designed to operate from a secure position (versus hand held nozzles) such as a platform atop a pumper, or when attached to an aerial ladder or elevating platform. They may be portable or permanently attached to the ap-

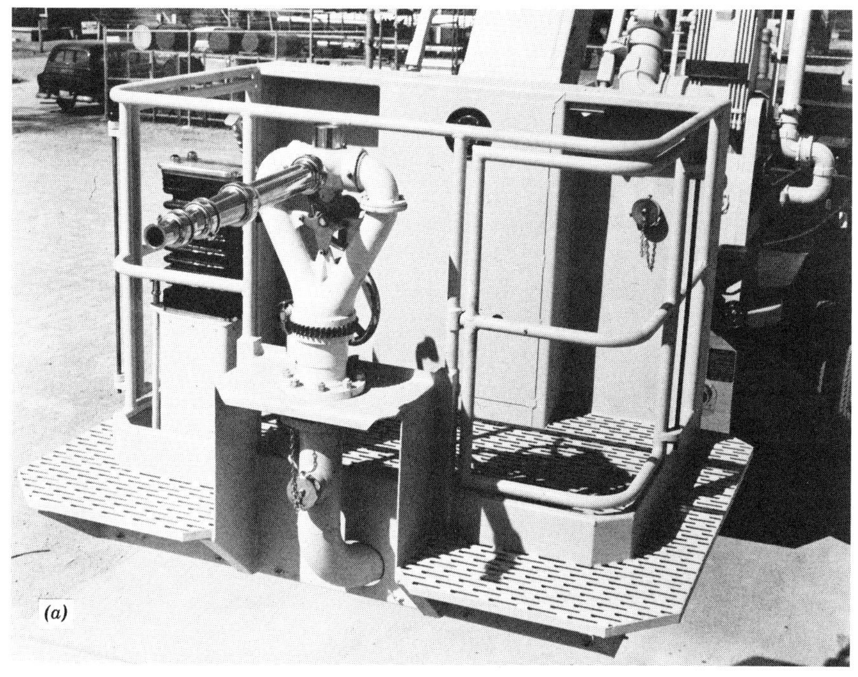

(b)

Fig. 5.16. Master stream appliances: *(a)* elevated platform (Courtesy Crown Coach Corp.), *(b)* mounted on the end of an extended aerial ladder. (Courtesy Akron Brass Co.)

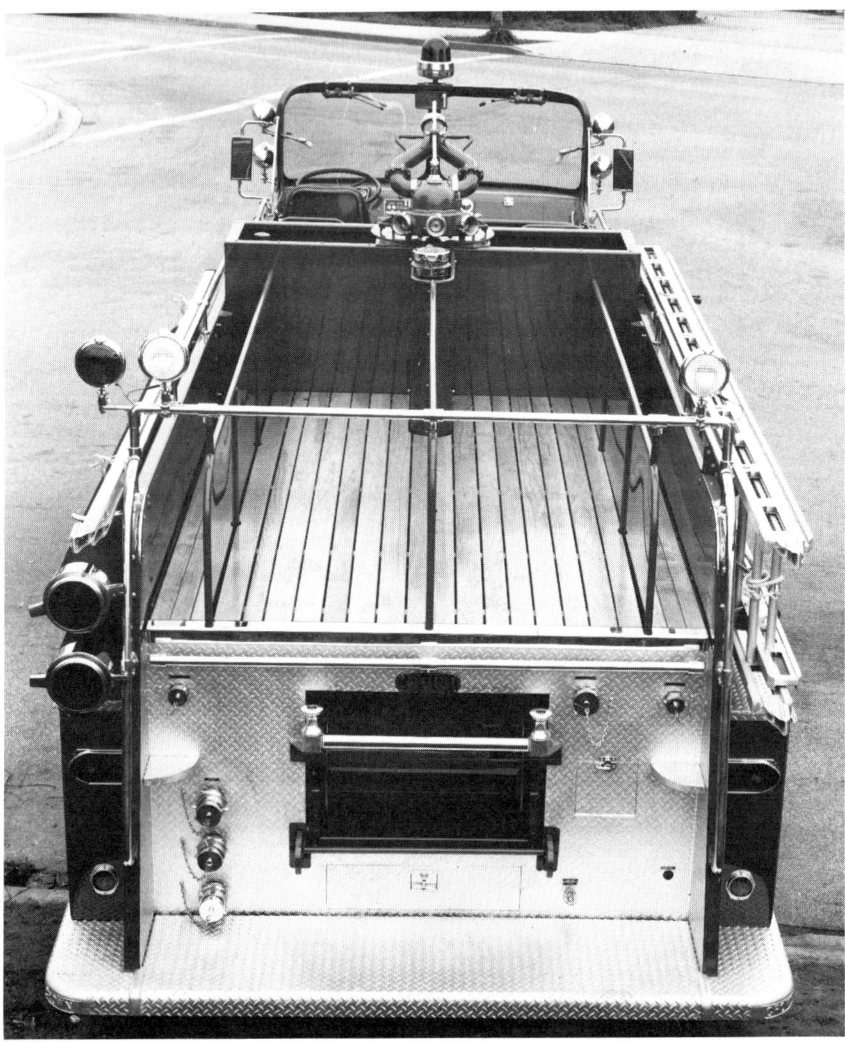

Fig. 5.16. Master stream appliances: *(c)* mounted on a pumper/engine. (Courtesy Crown Coach Corp.)

paratus (Fig. 5.16). Master stream nozzles are equipped to be supplied by several hoses or permanent piping, and have a selection of different large size tips (e.g., 1⅜ to 2¼ inches) and stream patterns, including combination streams. Master stream nozzles can also be remote controlled (e.g., elevating platforms) or manually controlled. The purpose of master stream nozzles is to provide large volumes of water to protect structures from radiant heat (with water curtains) and to provide cooling for large fires until they can be controlled by hand lines.

Now that we have some idea of the kinds of hose and accessories carried on apparatus, let's move to the subject of ladders.

Ladders

As noted in the section on apparatus, both the combination pumpers and other apparatus carry ladders referred to as ground ladders.

Ground ladders (Fig. 5.17) vary in construction, sizes, and uses. For example, solid or trussed beams may be used for the two parallel sides of the ladder. These beams may be of wood or metal (usually aluminum), or a combination of wood and metal; the same variations are found in the ladder rungs. Some ladders (extension ladders) will extend their length via pulleys and ropes; some are designed to collapse (attic ladders) to make more compact packages. The former type uses extensions (flys) that move up and down and are locked into position with metal catches (pawls) that fit over the rungs. The collapsible ladder is designed so the beams fold, making it easier to carry the ladder into a confined space. There are ladders with hooks that collapse against the beams until needed. These are called roof ladders, since the hooks are designed to hook over the roof peak of a building to hold the ladder in place.

All ground ladders, regardless of kind, require manpower to position them and hold them steady when in use; the longer the ladder, the more men it takes to move, raise, and hold them in position. With the increased use of aluminum ladders some of the manpower requirements have been reduced, but when in use, for minimum safety, they still require a person to hold the ladder steady (at the foot).

Ladders have uses other than helping one reach high places. Two examples are (1) using a ladder by laying it on a ice-covered pond or lake to aid in the rescue of a victim who has fallen through the ice; and (2) using three or more ladders set on their

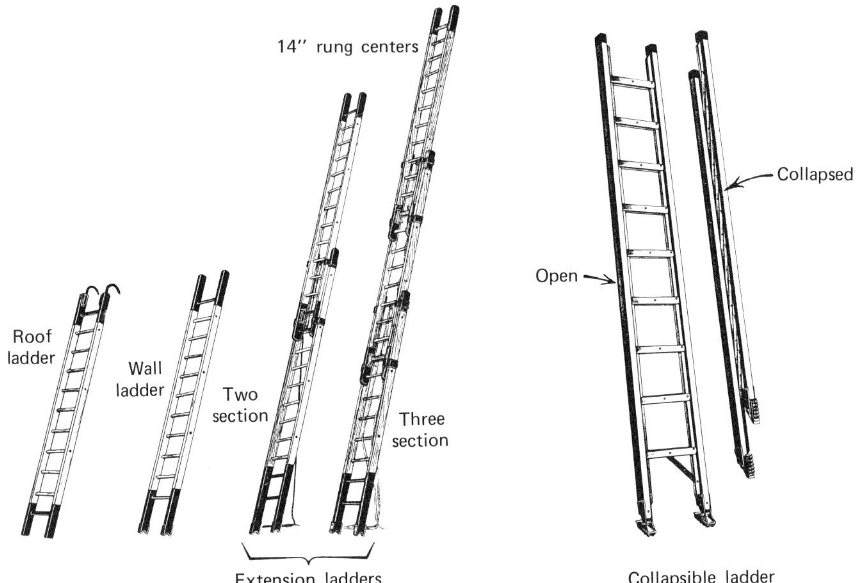

Fig. 5.17. Examples of ground ladders. (Courtesy Duo Safety Ladder Corp.)

beams and covered with salvage covers to frame a pool to hold an emergency water supply for fire fighting. Next we discuss other kinds of equipment, such as forcible entry tools.

Forcible Entry and Other Hand Tools

Within this category of equipment we include axes (pick-headed or fireman's, and flat-headed), prybars, crowbars, door openers, pikepoles, hydraulically operated tools, circular saws, hammers, and other tools that are used to gain entry into a vehicle or building. Various kinds and numbers of these items are carried on all pieces of apparatus; some kinds of apparatus carry more than others. For example, a rescue unit will carry many tools that will help lift, pry, and cut into metals and wood. A truck (aerial, elevating platform) will carry more forcible entry items than a pumper. However, "the bread and butter" forcible entry tool remains the axe, and it is carried by all fire department apparatus.

Salvage Equipment

For our purposes, salvage equipment includes salvage covers, smoke ejectors, shovels, hay or bale hooks, and similar items.

Salvage covers are just large pieces of waterproof canvas (Fig. 5.18) or plastic-coated nylon that are used to cover furniture, machinery, and other goods to prevent excessive water and soot damage during and after fire fighting operations. Salvage covers range in size from 12 feet by 12 feet to 14 feet by 18 feet. Salvage covers can also be used to drain off water from upper floors, as carrying bags to remove fire debris, and, when used in conjunction with ladders, to form an emergency pool of water, as mentioned previously. Most of the fire department's salvage covers are carried on salvage units and ladder trucks; usually, at least one or more will also be carried on each pumper.

Smoke ejectors also aid in salvage work; their proper use can reduce smoke damage in a building; and if explosion-proof, the ejector can be used to evacuate toxic and explosive gases from within a confined space. In essence, a smoke ejector is nothing more than a large portable fan that is used to push in or pull out air from an area. Smoke ejectors, depending on the design of their motor, can run on AC, DC, or AC/DC electrical current. They are designed to move large volumes of air quickly and reduces smoke damage by taking smoke via the fan from inside a building to outside where it can do no harm. Usually, at least one smoke ejector is carried on each combination pumper, and one or more on salvage and ladder truck apparatus.

Fig. 5.18. A folded salvage cover shown as it is carried on fire department apparatus. (Courtesy Globe Mfg. Co., Pittsfield, N.H.—Fire Suits & Safety Products.)

Other Equipment

To round out our general discussion of equipment we must mention extinguishers, rescue equipment, and water additives (foam, slippery water).

Rescue equipment includes a well-stocked first aid kit, at least one per piece of apparatus, and a resuscitator. Many departments carry a resuscitator on each pumper, as well as one on each special rescue apparatus. A resuscitator consists of small oxygen bottles with a regulator and various face masks (adult, child, infant); it is used to provide oxygen to a victim, either by inhalation or resuscitation (the application of rhythmic pressure which results in alternating positive and negative pressure in the lungs of the victim who cannot breathe on his own). Other rescue equipment includes a variety of splints, blankets, stretcher and/or litter, body bags, and other items. Paramedic rescue vehicles also carry more sophisticated equipment such as a defibrillator, electrocardiogram recorders, and intravenous solutions—to name just a few items.

Extinguishers are also carried on apparatus of all types. They must have extinguishing agents suitable for all types of fires; at least one dry chemical unit is usually carried on each vehicle. In addition, where grass and brush fire are possible, 5-gallon backpack pump-type extinguishers are included in the apparatus equipment. Other extinguishing aids, such as foam, are carried on combination apparatus to be used in special instances as when flammable liquid spills catch fire.

One other item of importance to fire fighters that may be classified either as rescue equipment or personal protection is the self-contained breathing apparatus (Fig. 5.19). This piece of equipment protects the individual fire fighter from superheated air, toxic gases, fire, and other dangers. The self-contained breathing unit consists of one bottle of compressed air (not straight oxygen) that may last for 30 minutes (depending on its design and the demand of the wearer), a mask, and regulator. Other breathing devices include the canister mask, which is being replaced by self-contained breathing equipment in many fire departments. This canister mask is similar to the "gas" mask formally utilized in war time. It has a canister of chemicals that purify the air, to a limited extent, of toxic gases; however, it cannot be used in all types of atmospheres, because it will not filter out all types of toxic materials. Therefore, the best protection is provided by the self-contained breathing equipment. Since we are discussing personal protection, let's move to the protective equipment used by fire fighters.

Personal Protection Equipment

A fire fighter is called upon to perform his job in all kinds of conditions, most of which are unsafe unless he has protective equipment. A fire fighter must work in a superheated atmosphere, toxic, corrosive, and explosive atmospheres, oxygen-depleted atmospheres and in freezing weather; he must also be prepared to handle radioactive material, pesticides, explosives, and other hazardous materials that are not conducive to life. He may also be subjected to situations where falling objects may strike him, floors may collapse beneath him, superheated water may drench him, and flames may lick about his clothing. Given these possible conditions, every part of the body should be protected. Here is how it is usually accomplished.

Head protection is provided by a helmet. The helmet may be of the traditional design or the more modern full-head protection design (Fig. 5.20). Each type of helmet has its proponents and opponents; therefore, the kind of helmet used varies with the fire department. Both helmets are designed to protect the head from blows and also to provide protection from hot water by channeling it away from the face and neck. The large rear overhang of the traditional helmet can be used for a face shield if the helmet is turned and worn with the back brim projecting forward over the face of the wearer. Face masks that flip into place on both types of helmets are also used for protection.

Body protection is provided by the turnout or bunker suit. (In special situations, such as those faced by airport crash crews, full exposure suits which differ from the standard turnout suit are used). A full set of turnouts consists of coat and pants, and, of course, suspenders. Both coat and pants are constructed to provide protection from heat and cold, and ventilated to prevent the

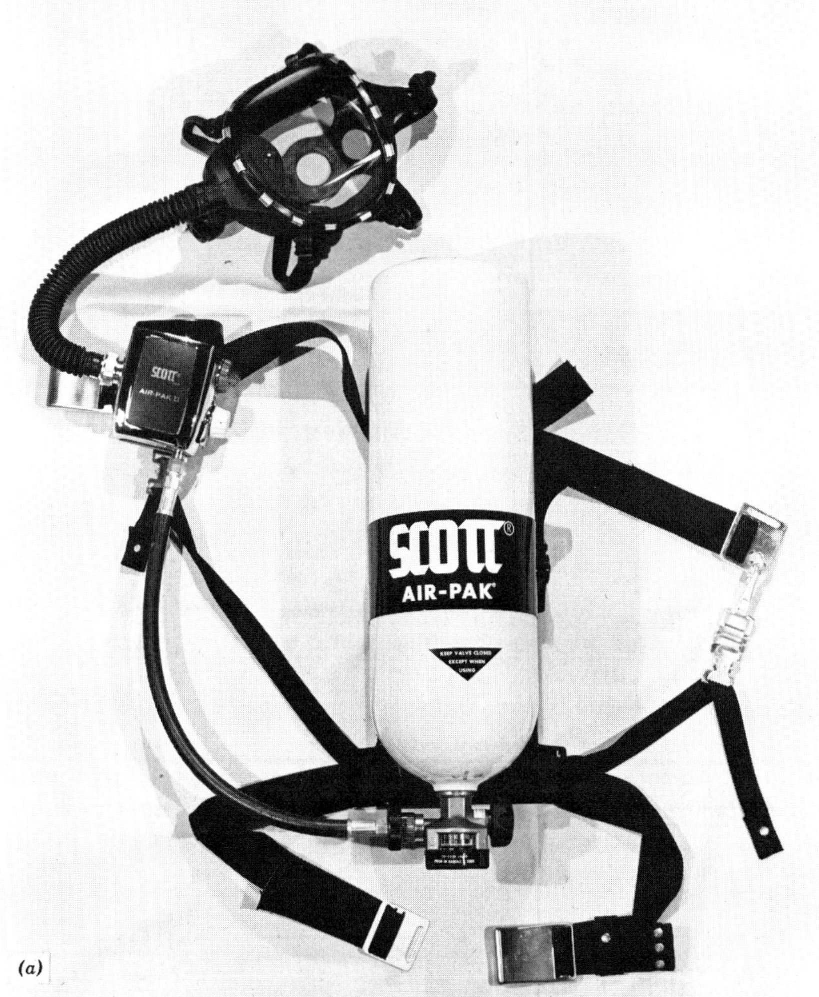

Fig. 5.19. *(a)* Self-contained breathing apparatus components.

body from overheating under fire fighting conditions (Fig. 5.21). Turnouts are made from several materials such as cotton duck, rubber, and flame-resistant fabrics such as Nomex. A standard feature of the coat is a large collar which helps protect the back of the neck. Snaps and other quick-action fasteners are used so that

Fig. 5.19. *(b)* As it is used by firefighters. (Courtesy of Scott Aviation, a division of A-T-O Inc., Lancaster, N.Y.)

Fig. 5.20. An example of the newer type of helmet; compare this style with the more traditional style shown in Fig. 5.19*b*. (Published by permission of the Federal Signal Corp., Signal Div.)

a minimum of time will be lost in putting on the coats and pants. During the day, the coats, helmets, and gloves may be the only turnout items used, whereas for night calls the full sets are used. (It often takes too much time to put on regular work boots and pants.)

The protection of feet is provided by thick rubber boots (turnout or bunker boots). The boots may be knee or hip length, and are designed to protect the feet (Fig. 5.22) from heat, cold, and caustic solutions, as well as from sharp objects such as nails and glass. They are designed to slip on and off quickly, yet, hopefully, to fit snugly enough to prevent blistering of the feet.

(a) Turnout bunker suit (b) Exposure/proximity suit

Fig. 5.21. Examples of protective clothing worn by firefighters. (*a*) The most common type of turnout or bunker suit. (*b*) Exposure suit similar to those used by aircraft crash truck personnel. (Courtesy Globe Mfg. Co., Pittsfield, N.H. Fire Suits–Safety Products.)

To complete the personal protection equipment we must include gloves (which must be worn at all times when working at a fire scene), flashlight, and hose spanner (Fig. 5.23). The spanner is used to tighten and loosen hose couplings and hydrant caps, and for similar uses; since it is usually made of brass, it can also be used for forcible entry in an emergency.

This concludes our discussion of apparatus and equipment; in the next chapter we will put these components together with others to give us a complete overview of the fire suppression system.

Fig. 5.22. Turnout or bunker boots. Note tread on the sole. (Courtesy Servus Rubber Co., Rubber Div.)

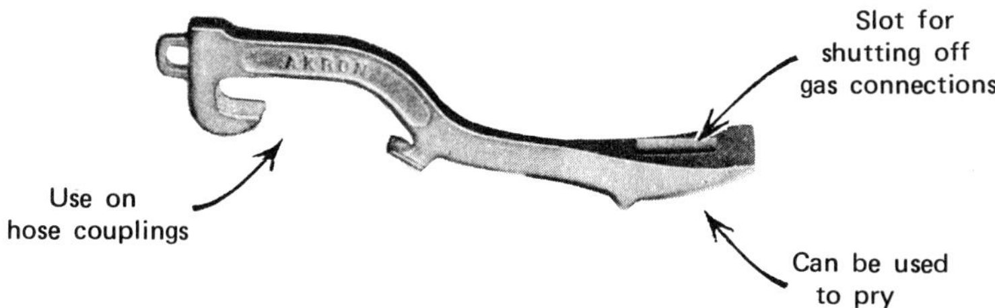

Fig. 5.23. A sample spanner that is usually carried in fire fighter turnouts. (Courtesy Akron Brass Co.)

SUMMARY

"Apparatus" and "equipment" refer a multitude of items utilized by a fire department. They include the motorized apparatus, equipment carried on the apparatus, and the protective clothing worn by fire fighters.

The basic piece of fire department apparatus is the pumper, or engine. It has the longest history of constant use in fire protection. The modern pumper/engine is or can be designed to do more than just pump water. It is a combination vehicle that carries hose, ladders, and its own water, in addition to other fire department equipment. It is common to refer to a pumper by the number of gallons the pump is rated to produce per minute and by how many functions the pumper can fulfill. Thus, a pumper may be referred to as a triple combination, 1000 gallon per minute pumper or engine. This tells us that it carries water, has its own hose and an assortment of ground ladders, and has a pump capable of 1000 gpm.

The basic pumper design can also be adapted to fulfill special duties such as wildland and airport fire protection. Furthermore, it can be designed to carry its own elevated hose stream in addition to the normal pumper equipment.

Tankers are another piece of apparatus that can be designed to fulfill one or more functions. Its basic function is to make a mobile water supply (1000 to 5000 or more gallons) available to other apparatus. Tankers are used in areas where other water supplies such as hydrants are not at hand. A tanker can also be designed to carry its own pump plus other equipment so that its function is broadened.

Aerial ladder trucks originally carried a wooden ladder, which was operated by a large spring and many men. The aerials of today use hydraulic systems to raise the long attached ladder and move it into a useable position. The basic function of aerials remains the same as that of its early forerunners—providing access to multistoried buildings. The aerial can also be designed as a multipurpose piece of apparatus, capable of pumping, providing elevated hose streams, and carrying additional equipment, including a selection of ground ladders.

An apparatus similar in purpose to the aerial is the elevating platform, or snorkel. The aerial and the elevating platform have different advantages and disadvantages, and their use depends on the attitude and finances of the particular fire department. Where financially possible, fire departments have both types of apparatus, which allows for greater flexibility in fire fighting operations.

A fire department may have other special types of apparatus to perform specific tasks such as portable lighting, rescue and/or paramedic, airport crash, and salvage operations. Where waterfront properties or shipping need fire protection, fireboats of various sizes are part of a department's apparatus. Aircraft are used where appropriate both to combat fire and to serve as transportation for men and equipment. All of the apparatus, including chief's cars, are equipped with warning devices and two-way radios to provide communication with each other and the alarm center and to facilitate rapid response to an emergency.

All major apparatus is constructed to specifications and standards set down by the fire department and the National Fire Protection Association. In addition, apparatus must meet the standards of the Department of Transportation and the Environmental Protection Agency. Each piece of apparatus must pass certification and acceptance tests prior to being used for fire protection.

Fire department equipment includes hoses, nozzles, and fittings. Most departments commonly use the double-jacketed rubber-lined hose, which may be of any size but is usually 1½ and 2½ inches in diameter. Other types of hose include ¾–inch booster/chemical hose, and hard and soft suction hose. Each type serves a particular purpose, and all are provided with male and female couplings of brass or metal alloy. Unlined hose is used both by public fire departments and on private fire protection systems such as standpipes. Recently, larger-diameter lined hose has been gaining wider acceptance for supply hose—in part because it reduces water power loss through friction.

Hose fittings such as the siamese and wyes are used to give greater flexibility to fire ground operations. Nozzles—another hose attachment—are varied in design so that they can give straight streams, fogs, or a combination of both with a flick of the wrist. Nozzle bases are designed to fit on the male coupling of hoses, and where larger master stream nozzles are required, several hoses are connected to the master stream appliance to give a greater volume of water to the tip.

Ground ladders, unlike the aerial ladders, are raised by manpower, and are carried on trucks, pumpers, and other fire department apparatus. Ladders of various lengths are available, and construction materials include wood and aluminum. Roof ladders and Collapsible ladders for use in small spaces and for gaining access via attic crawl holes are among the ladders designed for special functions.

Numerous forcible entry and hand tools are carried on the apparatus. Equipment such as salvage covers and smoke ejectors may also be carried. Other important kinds of equipment are resuscitators, extinguishers, first aid equipment, and water additives such as foam.

Personal safety equipment, which includes turnout or bunker suits, helmets, boots, and gloves, is also provided by the fire department. This equipment allows a fire fighter to gain some measure of safety while fire fighting. Self-contained breathing equipment allows a fire fighter to operate in an otherwise untenable atmosphere by providing a compressed, clean, air supply to his lungs on demand. All of the personal safety equipment, if used properly, will help protect the fire fighter in the performance of one of the most hazardous of the world's occupations—fire fighting.

REVIEW QUESTIONS

1. Define a pumper. List examples of equipment that may be carried on the apparatus.
2. How does an aerial ladder truck differ from an elevating platform?
3. What is the function of a tanker?
4. What is a tillerman? What is his function?
5. On what types of apparatus will you find the greatest number and selection of ground ladders?
6. How many types of ladders are carried on fire department apparatus? List them by name and explain the differences between each type.

7. In your opinion, why are aerial ladder trucks and elevating platforms necessary? What are their functions versus those of a pumper?

8. List the various kinds of apparatus, other than pumpers and trucks, that may be utilized by a fire department. Explain the function of each.

9. What are the tests that must be conducted before a new piece of apparatus may be placed in service for fire fighting? Why are they required? What standards must they meet?

10. What are baffles and where are they required? What is their purpose?

11. List the types of exterior coverings used on fire hose. What reduces friction in hose?

12. What is the function of suction hose?

13. What is the difference between a male and female coupling?

14. Where may unlined hose be utilized or found in the field?

15. What is the difference between a wye and a siamese?

16. List the kinds of nozzles that may be found in a fire department, and explain the different uses of each.

17. Can a master stream nozzle prevent a fire? How?

18. What are forcible entry tools? List some examples and how they may be used.

19. How can salvage covers be used in the field to accomplish different task?

20. List the uses of a smoke ejector.

21. What is the difference between a resuscitator and self-contained breathing equipment? How are they used?

22. List the design features of a helmet and define their purpose.

23. How do turnout or bunker suits protect fire fighters? Why is the protection needed?

REFERENCES

National Fire Protection Association, *Fire Protection Handbook,* 14th ed., Section 9, Boston, MA, 1976.

National Fire Protection Association, *National Fire Codes,* Vol. 2, Boston, MA, 1975.

chapter 6

6 FIRE SUPPRESSION

The ability of a fire department to put out fires and aid citizens in other emergencies requires a complex, efficient system. The system includes organization, station locations, stations, apparatus and equipment, training, communications, and manpower. Further, to do a more than adequate job of suppressing fires requires manpower versed in both the suppressing of fires and fire prevention.

In this chapter, our discussion will cover the major components of the fire suppression system and those components needed for notification of fire, apparatus, response, and suppression. Thus, at the end of this chapter we should have a complete overview of the fire suppression division of a fire department.

At the end of this chapter you will be able to:

1. Identify the factors involved in the location of fire stations and the impact of their location on response time.
2. Identify the ranks and duties of personnel assigned to a fire station.
3. Identify the important aspects of training and the need for a training center.
4. Identify the methods of communication utilized by a fire department and their purpose.

5. Identify the basic components in a municipal alarm system and how an alarm is received by a fire company.

New Terms

Response time	*Battalion chief*
Apparatus floor	*Multiple alarm*
Ready or day room	*District chief*
First-in or first-due	*Radio code*
Engine company	*Street alarm*
Truck company	*Company run or alarm Report*
Company officer	*Alarm and communications center*
Driver-engineer	

The purchase of the apparatus and equipment described in the last chapter is just the beginning of fire suppression. The apparatus, equipment, manpower, communications, and alarm systems must be appropriately located and housed to provide the quickest and most efficient response to any emergency that may require fire department assistance. An alarm or telephone call for assistance must be received by someone before any action can be taken. Any action must be controlled and directed with a minimum of lost time. Seconds, not minutes, become important. Time cannot be lost while someone finds proper equipment and locates others who can use the equipment. Everything must be integrated to work together to form an efficient system, and the base for that system is the fire station.

FIRE STATIONS

The fire station is the center of all fire protection activity, shelter for the expensive apparatus, and home for the fire fighters. Looking at a fire station from the exterior in any given city or town, one might think it almost deserted; it appears very quiet and unobtrusive until an alarm sounds. Suddenly doors rattle open, engines roar, the occupants inside scurry about, a siren screams, air horns blast their warnings, and in a roar of sound the apparatus flashes down the street, leaving the station silent and abandoned until its occupants return or others come to take their place.

In a fire department with well-trained personnel, the time elapsed between the receipt of an alarm and the departure of the apparatus is 30 seconds or less. If the station is located properly, the apparatus will be at the scene of the emergency within three minutes or less. It seems amazing, but the design makes it possible. Much thought and planning, and trial and error experience, have refined the choosing of station location, station design, and design of auxiliary systems such as emergency notification, and the determination of number of men, and method of manning. It is not by chance that response time is cut to the minimum. Everything possible is done to leave little to chance in fire protection. For example, the traditional brass pole, in stations with upper floors, is still in use because it remains the fastest device for moving people safely and quickly to a lower floor. But other than having a brass pole, what is a station like? What kinds are there? How do they function?

For a small fire department, one station may be enough to fill the fire suppression needs of a community. If there is only one station, it serves a combination of needs and becomes the heart and brain of the community's fire protection system. The one station is proclaimed the "main" station or headquarters station. It serves as the administrative office of the chief and fire prevention bureau, training center, and alarm and dispatch center. It may also function as a location for registration of voters, casting voting, issuing bicycle licenses and burning permits, and a myriad of other community services which are limited only by the chief of the department and the political body of the city or town. The station is usually located near the center of town, near the business district, in order to protect the high-value area of businesses. In areas where suburbs have developed and the core downtown areas are dying, this central location of stations may be altered.

The interior of the station (Fig. 6.1) contains the apparatus floor, storage for hose and other equipment, alarm center, and

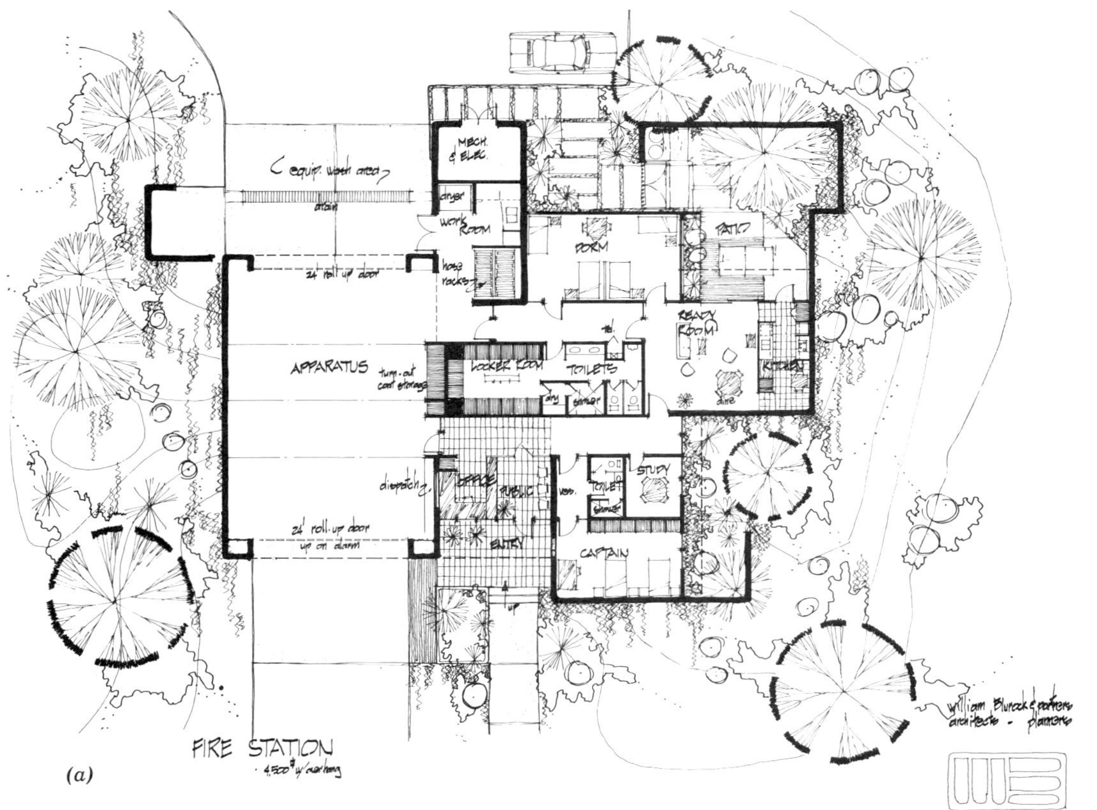

(a)

Fig. 6.1. *(a)* A floor plan of a typical single-story fire station. (Courtesy City of Huntington Beach, & Wm. Blurock, Architect.)

Fig. 6.1. *(b)* A typical drill or classroom. (Courtesy City of Orange FD—G. Conniff.)

Fig. 6.1. *(c)* Ready or day room in a fire station. (Courtesy City of Orange FD—G. Conniff.)

Fig. 6.1. *(d)* A typical apparatus floor. Note the tractor trailer aerial with the tillerman position at the rear of the trailer. (Courtesy City of Orange FD—G. Conniff.)

administrative offices. In addition, there is a kitchen for the use of the fire fighters, a dormitory for those on shift, and a ready or day room used both as a classroom and for relaxation. In one-story stations and in those that are not headquarters (in areas where there are more than one station) the offices are not present, and the station layout is usually simpler.

If more than one station is within the department's jurisdiction, station location is determined by a combination of factors. Even in a small town, more than one station may be required because of the physical layout of the town. In other words, man-made or natural obstacles, or a combination of both, may make another station or stations essential. For example, rivers, railroad tracks, freeways, and dead-end streets may make additional stations necessary. For purposes of illustration, let's take a small town that is divided by a main track of a railroad. It is a busy track, with freight trains of various lengths using it frequently. There is a possibility that a fire could occur across the tracks from the fire station and apparatus would be blocked by a train passing through so that a critical time-wasting detour would be required. To prevent such an occurrence, another station on the other side of the tracks would be required to insure fire protection for both sides of the town at all times. In major cities these same obstructions are present, and they also determine the station locations in large departments. Other factors such as distance of travel and types of buildings and their uses play a determining role in the location and number of stations a fire department may have.

Where a fire department has more than one station, the fire protection of the city is broken into areas of primary responsibility and units of coverage per station. Each station is assigned a geographical area in which it is first-in or first-due. In other words, the first-due station is the one to arrive first on the scene of a fire or other emergency reported within its geographical district. If a large fire occurs in another station's area additional apparatus is moved in to aid the first-due, and other apparatus is moved from other parts of the city into the vacated stations. If all the fire department's equipment is busy at the scene, then apparatus from nearby fire departments will be called in to cover the stations vacated in the city. In essence then, at no time is a portion of a city left without any fire protection; apparatus is always moved up or outside aid comes from other fire departments to cover an unprotected area.

Manning

Within each station in a fire department you will find the smallest unit of fire fighting: the company. The term company refers to the men assigned to the station and to a particular piece of apparatus. For example, those stations with a single pumper and a pumper crew are referred to as engine companies. If an aerial or elevating platform is in a station, it and its crew are referred to as a truck company. If a station has both a pumper and an aerial, then it contains an engine company and a truck company. Other special pieces of apparatus and men will usually be termed squads; thus, we speak of the rescue squad.

Each engine, truck, or other piece of apparatus is under the control of a company officer, who will have, depending on the fire department, the rank of captain or lieutenant. Under the company officer, there is the driver-engineer of each piece of apparatus.

The driver-engineer is responsible for the care of the apparatus (under the direction of the company officer), for driving the apparatus to a fire, and for operation of the apparatus (e.g., pumping operations, aerial operation). In most departments, the driver-engineer is ranked above a fire fighter and just below the company officer.

Within the general designation of fire fighter a fire department may make various distinctions. For example, some departments assign specific duties, so a fire fighter may be a nozzleman, hydrantman, or if qualified, a paramedic. If a fire fighter's duties entail activities such as raising ladders, ventilating buildings, and other duties assigned specifically to a truck company he may be called a truckman.

In overall charge of a specific station you will find the highest ranking officer, who is usually a captain. He is responsible for the actions of the companies assigned to the station and the overall operation of station activities. He is also in charge at the scene of an emergency until a senior captain or higher ranking officer arrives and takes command.

In departments with several stations and districts, the next rank above the captain will be either a battalion or district chief (the lowest ranking chief officer). The battalion chief is responsible for several stations within the department. Therefore, if there is more than one district of several stations a chief officer will be assigned to each group of stations. Above this level are those ranks we discussed in Chapter 4 (e.g., assistant chief, deputy chief, chief of department).

Regardless of the number of stations within a department or number and kinds of ranks, the fire fighting activities would not be worth much without standardized training, teamwork, communications, and systems to notify the personnal of fires and other emergencies. Let's examine these other components necessary for fire suppression, beginning with training.

TRAINING

In small departments, both volunteer and paid, the training is usually handled by a company officer assigned the responsibility by the chief. The training is usually spasmodic due to the other regular duties that must be performed by the officer, such as station maintenance, operations, and communications. The training facilities are usually limited to the confines of the apparatus floor, recreation room, and yard area; city streets are used when possible. In brief, training is limited due to lack of proper facilities and other restrictions.

In larger departments and those with sufficient funds, a training center is available. Here the training includes courses for new fire fighters and refresher courses for companies and company operations. An example of an excellent training facility is shown in Figure 6.2. This center was constructed with funds provided by more than one department through a joint funding agreement. Note that there are mock streets, mock props for flammable liquid and gas fires, a tower to simulate multistoried buildings (complete with standpipe and sprinkler systems), and a building housing the communications center, classrooms, areas for multimedia presentations, and offices for training personnel. In this instance, the training program is under charge of a chief officer, and company officers and fire fighters are assigned as their primary duty, the task of providing a full training program.

In this kind of training center fires are built, and joint operations of truck and engine companies are carried out under realistic conditions, which obviously aids in developing efficient fire fighting operations (Fig. 6.3). Training is also carried out at the station by each company officer to insure that all fire fighters and officers are well prepared to handle their jobs. There can never be too much training, and learning must go on continually, regardless of length of service in a fire department. Efficient fire fighting is a result of training, teamwork, and experience, and a center such as

a
b
c
d
e
f

Fig. 6.3. A typical multicompany drill at a training center. (Courtesy Crown Coach Corp.)

the one described here is essential for the upgrading of fire fighting operations.

We next consider communications.

Communications

No amount of training would be useful, without communications. Training itself relies on both verbal and written communications.

Fig. 6.2. An example of a complete fire department training center. (*a*) Oil and flammable liquid-fire mockups. (*b*) Control and observation tower. (*c*) Rescue practice area. Note vehicles. (*d*) Training and drill tower with interior and exterior stairs, standpipe, automatic sprinklers, and rooms used for training fires. (*e*) Fire station. (*f*) Training and communications center with classrooms. (Courtesy Huntington Beach FD.)

Written communications include records and reports. Some examples of the records and reports are company run reports, hose records, apparatus records, monthly station reports and daily logs, inspection and prefire planning reports, and personnel

Entries contained in this report are intended for the sole use of the State Fire Marshal. Estimations and evaluations made herein represent "most likely" and "most probable" cause and effect. Any representation as to the validity or accuracy of reported conditions outside the State Fire Marshal's Office, is neither intended nor implied.

STATE OF CALIFORNIA
OFFICE OF THE STATE FIRE MARSHAL
FIRE INCIDENT REPORT

INCIDENT NO.

DEL 1 ☐ CORR 2 ☐

________________ FIRE DEPARTMENT
(DEPARTMENTAL USE)

1 OCCUPANT NAME | RELATIONSHIP | ALARM SOURCE: TEL. ☐ PFAS ☐ RADIO ☐ BOX ☐ VERBAL ☐ OTHER ☐

2 ADDRESS | ROOM / APT. NO. | CITY | ZIP | TELEPHONE NO. (CALL BACK)

3 OWNER NAME | ADDRESS | CITY | ZIP | CENSUS/PARCEL NO.

4 MANAGER NAME | ADDRESS | CITY | ZIP | TELEPHONE NO.

A. INFORMATION (PAGE)

1 FIRE DEPT. ID | INCIDENT NO. | EXPOSURE NO. | TIME | MONTH | DAY | YEAR | DAY CODE | COUNTY OF FIRE | DIST/CITY | OUT OF JURISDICTION CHECK IF YES ☐

B. PROPERTY CLASSIFICATION (PAGE)

1 CODE | TYPE OF INCIDENT | CONSTR. DATE PRE 72 1 ☐ POST 71 2 ☐

2 CODE | PROPERTY CLASSIFICATION (COMPLEX)

3 CODE | PROPERTY CLASSIFICATION (INDIVIDUAL)

C. PROPERTY TYPE (PAGE)

1 PROPERTY MANAGEMENT: PVT 1 ☐ FED 2 ☐ STATE 3 ☐ COUNTY 4 ☐ CITY 5 ☐ DISTRICT 6 ☐ FOREIGN 7 ☐ OTHER 8 ☐

2 CODE STRUCTURE, BUILDING OR VEHICLE - PROPERTY TYPE | BUILDING NO. STORIES

3 STRUCTURE, BUILDING OR VEHICLE - CONSTRUCTION TYPE: EXT. WALL N/C 1 ☐ COMB 2 ☐ | INT. WALL N/C 3 ☐ COMB 4 ☐ | FLOOR - ROOF N/C 5 ☐ COMB 6 ☐ | FIRE RATED YES 7 ☐ NO 8 ☐

D. EXTENT OF DAMAGE (PAGE)

1 CODE | EXTENT OF DAMAGE - FIRE

2 CODE | EXTENT OF DAMAGE - SMOKE

3 CODE | EXTENT OF DAMAGE - WATER

4 ESTIMATED LOSS - PROPERTY | ESTIMATED LOSS - CONTENTS

E. LOCATION & CAUSE (PAGE)

1 CODE | LEVEL OF ORIGIN

2 CODE | SOURCE OF HEAT CAUSING IGNITION

3 CODE | FORM OF HEAT CAUSING IGNITION

4 CODE | ACT OR OMISSION CAUSING IGNITION

F. AREA, MATERIALS & SMOKE SPREAD (PAGE)

1 CODE | AREA OF ORIGIN

2 CODE | TYPE OF MATERIAL FIRST IGNITED

3 CODE | FORM OF MATERIAL FIRST IGNITED

4 CODE | MAIN AVENUES SMOKE SPREAD

G. SPREAD OF FIRE (PAGE)

1 CODE | MAIN AVENUES FIRE SPREAD

2 CODE | TYPE MATERIAL CAUSING SPREAD

3 CODE | FORM MATERIAL CAUSING SPREAD

4 CODE | ACT OR OMISSION CAUSING SPREAD

H. PROTECTION FACILITIES (PAGE)

1 CODE | SPRINKLERS - TYPE

2 CODE | SPRINKLERS - EFFECTIVENESS

3 CODE | STANDPIPES - TYPE

4 CODE | STANDPIPES - EFFECTIVENESS

5 CODE | PORTABLE EXTINGUISHERS - TYPE

6 CODE | PORTABLE EXTINGUISHERS - EFFECTIVENESS

I. PROTECTION FACILITIES (PAGE)

1 CODE | PRIVATE BRIGADE - TYPE

2 CODE | PRIVATE BRIGADE - EFFECTIVENESS

3 CODE | SPECIAL HAZARD PROTECTION - TYPE

4 CODE | SPECIAL HAZARD PROTECTION - EFFECTIVENESS

5 CODE | SIGNAL OR WARNING SYSTEM TYPE | CODE | EFFECTIVENESS

6 CODE | SIGNAL WARNING SYSTEM - MEANS OF ACTIVATION

7 CODE | SIGNAL/WARNING SYSTEM - TYPE DETECTORS

CODE | WATCHMAN EFFECTIVENESS | CODE | OTHER FACILITIES EFFECTIVENESS

J. MISCELLANEOUS (PAGE)

1 FIREFIGHTER: NO. INJURED | NO. OF DEATHS | CIVILIANS: NO. INJURED | NO. OF DEATHS

2 SFM FORM 60-1 SUBMITTED FOR EACH DEATH (CHECK BOX IF YES) ☐

SFM FORM 60-60 (7/73)

performance reports. Each report has a specific function and describes a small part of the fire suppression division's activities and needs. For example, the company run report is a written record (Fig. 6.4) of each call that the company has responded to. It tells the reader what kind of call it was (e.g., fire, rescue), what actions

K. RESCUE OR INHALATOR

NAME OF PATIENT	AGE	SEX	TELEPHONE NO.
ADDRESS	PARENT OR GUARDIAN		
CONDITION ON ARRIVAL GOOD ☐ FAIR ☐ POOR ☐	CONDITION ON DEPARTURE GOOD ☐ FAIR ☐ POOR ☐	CAUSE OR AILMENT	
AID GIVEN INHALATOR ☐ HRS ____ MIN ____ RESUSCITATOR ☐ HRS ____ MIN ____ OTHER ____			
AMBULANCE	PHYSICIAN	TAKEN TO	

L. FIREFIGHTING APPARATUS

COMPANY	RESPONSE TIME	SHUT-OFF TIME	OFFICERS	FIREMEN			PUMPERS RESPONDING	PUMPER CREW WORKED	TRUCKS RESPONDING	TRUCK CREWS WORKED
				ON DUTY	OFF DUTY	VOL.				

EQUIPMENT LOST OR DAMAGED	EQUIPMENT NEEDED BUT NOT AVAILABLE

M. HOSE & LADDERS

HOSE LINES	1"	1½"	2½"	OVER 2½"	BATTERY	LADDERS
NO. OF LINES						
NO. OF FEET						

N. MISCELLANEOUS INFORMATION

WETTING AGENT GALS.	FOAM GALS.	WATER USED GALS.	WEATHER COND.
MUTUAL AID	POLICE UNITS	FIRE PREVENTION	PHOTOS BY
IF EQUIPMENT INVOLVED IN IGNITION TYPE	YEAR	MAKE	MODEL
INSURANCE COMPANY NAME OR AGENT	PHONE	AMOUNT OF COVERAGE CONTENT	PROPERTY
INSURANCE COMPANY ADJUSTER	PHONE	POLICY NO.	
ESTIMATED VALUE CONTENT	PROPERTY	ACTUAL LOSS CONTENT	PROPERTY

O. VEHICLE

TYPE OF VEHICLE	MAKE	MODEL	YEAR	LICENSE NO.	STATE
REGISTERED OWNER	ADDRESS		TELEPHONE NO.		

Entries contained in this report are intended for the sole use of the State Fire Marshal. Estimations and evaluations made herein represent "most likely" and "most probable" cause and effect. Any representation as to the validity or accuracy of reported conditions outside the State Fire Marshal's office, is neither intended nor implied.

SIGNATURES

PREPARED BY	DATE
APPROVED BY	DATE

Fig. 6.4. A company run or fire incident report designed for use with a code book and suitable for data processing use. (California State Fire Marshal's Office.)

were taken by the company, and the results of those actions. This report is used as a record of the fires or other emergencies the department responds to daily.

A list of some written reports and their functions follows:

Hose records	These describe the condition and location of each length of hose in the department. Individual records for each length of hose may also be kept.
Apparatus records	There are basically two kinds: (1) A checklist used daily to show the condition of each piece of apparatus that is normally used by the engineer-driver assigned to that particular piece of apparatus. (2) The record of mechanical repair and service, as well as of pump and ladder tests which is kept by the mechanics who maintain the apparatus.
Personnel performance reports	These are reviews of the performance of each department employee. They are usually filled out at least once a year and before every pay raise or promotion.

The communication system of a fire department includes the use of two-way radio. The minimum radio communication system uses a base station and radios in each piece of apparatus and all other mobile units of the fire department. It is a rare department that does not utilize radio. Radio permits rapid communication during normal day to day activities and in the event of an emergency. The base station may be manned by fire fighters or by civilian alarm communications operators, who may or may not be part of the fire department. In some areas the radio communications are handled by communication centers that also handle police activity and other services of the city or county that utilize radio equipment.

To prevent or minimize misunderstanding of radio communications, various numbers are used to indicate the basic message being transmitted. One example is the 10 code; samples of the code are shown in Table 6.1. Note that at the top of the table additional meanings are given; for example "Code 3" indicates how to respond from your location to the scene of emergency. As is evident, the use of the proper code number can convey the message quickly and with the least chance of misunderstanding. Memorization of the radio code used by the department is one of the first jobs that a new fire fighter will be called upon to do.

Our next subject, alarm systems, can be considered part of communications, but for our purposes it is best discussed under a separate heading.

ALARM SYSTEMS

The telephone and the street alarm boxes bear the brunt of the work of receiving and transmitting alarms of emergencies and

TABLE 6.1.
Uniform State Fire Mutual Aid Radio Code

CODE 1	At your convenience	CODE 3	With red light and siren
CODE 2	Without siren or red light	CODE 4	No further assistance needed

CODE NO.	Meaning	CODE NO.	Meaning
10-1	Reception of your transmission is poor, scratchy, low volume, or unreadable	10-9	Repeat your message
10-2	Reception of your transmission is good, loud, and clear	10-19	Returning to station or return as directed
10-4	Your message received and understood and complied with, or affirmative	10-20	What is your location?
10-7	This unit is out of service	10-97	This unit has arrived at the location
10-8	This unit is now in service	10-98	Finished with assignment

other information regarding fire department operations. When telephones were not as common as they are now, the street corner alarm box was the workhorse. Now the street alarm boxes serve a limited function (and are the source of the majority of false alarms) and are being replaced by the telephone and telephone-type equipment. However, in many departments the street alarm box is still used and deserves our attention.

The individual street corner alarm box is a part of a network of alarm boxes within a city. Each box is tied into the main receiving station and has its own identification number indicating a specific location. With the exception of the newer radio transmitter boxes, all the boxes rely on overhead or underground electrical power lines for operation and to tie it to the total system.

The function of the street box (other kinds of private alarm systems will be discussed in a later chapter) is to notify the fire department that the box has been tripped or pulled (the basic operation is similar to tripping a light switch off and on). Each alarm box has its own number on a coded wheel, and when the box is pulled it sends a series of coded electrical impulses to the central alarm receiving station.

As the pulses reach the receiving station they are recorded visually and audibly, both in the receiving station and in all fire stations in the department. The visual record is usually made on a paper tape either by inked marks or by actual punches in the tape. The aural notification is usually a bell, gong, or other noisemaking device. In addition within each station selected lights usually go on when an alarm comes in. At night, when an alarm comes in, the station crew is rudely awakened by a ringing bell and bright lights. Parts of the same system are generally used to transmit (actually retransmit) alarms received by telephone, so that bells or other devices sound in the fire stations, notifying everyone that an alarm has been received.

In addition to the alarm provided by the street boxes and auxiliary systems, radio, teletype, and/or private phone lines are used to give specific information about an alarm to the fire suppression crews. This is usually handled by those who are assigned duties in the alarm or alarm and communications center. In essence, the purpose of the alarm system is to give as much factual information as possible regarding the location, type of emergency, cross streets, nearest hydrants and so forth to those responding to the alarm. Further information as to the apparatus assignments and moveup into vacated stations is handled using the same methods. Each time more apparatus is required at the scene of an emergency, another alarm is put in; from this system we have the terminology of two-alarm, three-alarm fire which indicates that it took more than just the first-due apparatus to control the emergency.

In our chapter entitled "Tactics and Strategy" we will see how the suppression system functions after the alarm is received. However, the part of the fire protection system to be discussed in the next chapter is the fire prevention efforts of a fire department.

SUMMARY

The fire station, a fire fighter's home away from home, serves as the center of the fire suppression system. The location of the fire station is critical to fast response to any emergency that may occur within a jurisdiction. Well-located stations plus good training will reduce the response time to a minimum, which means less life and property loss due to fire. Fire stations, by design, are utilitarian and provide all the necessary features to house both fire fighters and equipment with a minimum of wasted space and expenditure.

Within each station we find one or more companies, depending on the number and types of apparatus housed within that particular station (e.g., engine company, engine and truck companies). Each station is under the charge of a company officer—one per shift—who is responsible for its operation.

In many fire departments, the rank of the driver-engineer is used to designate those fire fighters who are responsible for the operation of the apparatus and routine maintenance of the apparatus's equipment. Where this rank is not used, experienced fire fighters are assigned to drive each piece of apparatus at the beginning of each shift. Fire fighters may also be given titles that refer to special tasks such as truckman, nozzleman, hoseman, hydrantman, and paramedic.

Where more than one fire station is required within the department's jurisdiction, other chief officer ranks may be used. For example, the rank of battalion chief or district chief may be used; these officers are responsible for several stations within their district; the company officers are then responsible directly to the battalion or district chief in charge.

Regardless of how new the apparatus and the number of fire fighters available, a fire department is only as effective as its training program. Even occasional training sessions conducted by company officers are better than no training, but ideally the department has an assigned training officer and a complete training facility.

Communication is the glue that bonds human endeavors together for a common purpose, and both written and oral communications are used by a fire department. Written communications are required for recording information concerned with alarm responses, hose, apparatus maintenance and testing, prevention inspections, and prefire planning, to name a few areas of importance. All of these records taken together provide a basis for budgeting and establishing the overall goals and objectives for a fire department.

Radio communication is an important part of fire ground operations. To minimize misunderstanding during radio communications, number codes have been devised to communicate basic message phrases (e.g., 10–4 means "message received and understood"). The use of such codes also reduces the cluttering of the radio frequency with unnecessary verbiage.

Alarms from street alarm boxes, telephone calls, and teletype communications from a remote alarm center may be received by a fire department in several ways. As the alarm is received by the alarm center and/or headquarters station, it is transmitted both to the station that is first-due and the rest of the stations in the department. From that point the fire suppression operations begin to take place.

REVIEW QUESTIONS

1. List those items that must be considered for the locations of a fire station. Note the significance of each item listed.
2. What is meant by response time?
3. Explain the term first-in.
4. If all the apparatus of a department is out on a fire call, how are the other areas of a city provided with fire protection?

5. Define a company. Who is in charge? What are his responsibilities?
6. What are the duties of a battalion chief?
7. Who is responsible for company training? In your opinion, what would be the kinds of training needed by a new fire fighter?
8. What types of communication are used by a fire department? List examples of each type and their purpose.
9. What is the purpose of a "10" code?
10. List the main points of how a street alarm box functions.
11. What kinds of information are given to a fire suppression crew by the alarm operator?
12. What is meant by a multiple-alarm fire?

REFERENCES

National Fire Protection Association, *Fire Protection Handbook,* Sections 9 and 12, Boston, MA, 1976.

National Fire Protection Association, National Fire Codes, Volume 7, Boston, MA, 1975.

International City Manager's Association, *Municipal Fire Administration,* Chapters 6–8, Chicago, IL, 1967.

chapter 7

7 FIRE PREVENTION

A good program of fire prevention is the fire department's first line of defense in its struggle against loss of life and property by fire. The effectiveness of a fire prevention program is dependent upon total fire department involvement; the program cannot successfully be relegated solely to those who are assigned to specific fire prevention duties (e.g., fire marshal, inspector). The smaller the number of fire department personnel working on fire prevention problems, the more the fire department will have to rely on its fire suppression activities to control the local fire problem.

Because manpower, or lack of it, affects the amount of fire prevention activities that are carried out, the attitude of the chief of the fire department toward fire prevention activities has maximum impact. If the chief feels that fire prevention is not of primary importance, then the effectiveness of the program will be severely curtailed. Ideally, in a well-balanced fire protection effort, specialists will be permanently assigned to fire prevention, and the balance of fire department manpower will be utilized to conduct routine inspections and help in the public education program as well as serve in fire suppression.

In this chapter, we will explore the essence of a fire prevention effort in a municipal fire department. In doing so, we will discuss the organization of a fire prevention bureau (or division), personnel requirements, and its relationship to the total fire department organization.

Further, we will explore some of the bureau's basic functions and responsibilities, such as inspection, enforcement, and education. In addition, we will also discuss a typical company fire inspection program, its functions, and operation.

At the end of the chapter, we will have accomplished the following objectives:

1. Identified who is responsible for a fire prevention program.
2. Identified who is responsible for the running of a fire prevention bureau and where his authority comes from.
3. Identified the objectives and functions of a fire prevention bureau.
4. Identified the basic qualifications necessary to perform the duties of a fire prevention officer.
5. Outlined a sample company fire inspection program.

New Terms

Fire marshal	*Inspection*
Inspector	*Citation*
Plan checking	*Notice of violation*
Company fire inspection	*Investigation*
Enforcement	*Weed control*

First let's determine the place of fire prevention in the total organization of a fire department. In a small fire department it may be one of two divisions: fire prevention and fire suppression. In a larger department it may be one of several divisions that include research and development, personnel, training, communications, fire suppression, and fire prevention. In both instances, the head of the fire prevention bureau is directly responsible to either the assistant chief or the chief of the fire department. In other words, the head of fire prevention is an important position. Now that we know where the fire prevention bureau fits in the total organization, the next thing to be determined is who is responsible for its activities.

WHO'S RESPONSIBLE?

Taking first things first, the final responsibility for the fire prevention program and its activities belongs to the chief of the fire department. However, if the chief of the department tried to do everything himself, it is unlikely that much would ever really be accomplished. Therefore, the chief delegates some of his responsibilities and duties to other members of the fire department. He delegates some authority to the head of the Fire Prevention bureau, who may be titled fire marshal. The title as well as the rank of this person varies from department to department (chief, captain, lieutenant); however, for our purposes we will refer to him as the fire marshal as the fire codes do.

The legal authority for the chief of the department to assign and delegate his responsibility to the fire marshal is found in both the fire code and in the rules and regulations of the fire department. In addition, th chief of the department, via the fire code, has the authority to assign any number of men, regardless of rank, to perform duties related to the enforcement of the fire prevention regulations and codes. In actual practice, the bulk of positions in fire prevention, including fire marshal, are filled via competitive civil service promotional examinations rather than by appointment without competition. Regarding the number of men who aid in fire prevention enforcement (e.g., company inspections), the chief normally makes a broad policy statement to the effect that every person in the fire department will aid in the fire prevention effort and in conducting a fire prevention program. So far, so good. Let's now proceed to the purpose of a fire prevention bureau.

PURPOSE

The purpose of a fire prevention Bureau is essentially the same as the purpose of total fire department: to protect life and property from fire. However, the methods by which the objectives are accomplished by a fire prevention division are different than those used by a fire suppression division. (In reality there should be no such distinction because every section of the fire department should be trying to achieve the same goal, using any and all methods.)

In fire prevention the objectives are achieved through preventing fires before they start and minimizing fire loss, in the event of fire, through good construction, adequate exits, built-in fire protection systems, enforcement, and public education. In fire suppression, theoretically, you prepare for fire by training manpower for combat, buying emergency equipment, prefire planning, and then trying to minimize fire loss once the fire has started. As can readily be seen, the two methods are compatible, necessary, and overlap, so that no one can claim that fire prevention is not part of his job as a fire fighter. To repeat, fire prevention is the first line of defense in providing fire protection. Actual fire fighting occurs when the defense has been broken; then confrontation and battle must take place on the fire's terms. Once the fire starts, the destruction is incessant and requires massive retaliation against an unchecked enemy, whereas fire prevention deals with saving without the initial destruction. Having discussed the objectives of fire prevention, let's now determine how a fire prevention program moves to meet the objectives.

FUNCTIONS

As in any situation in which we strive to reach goals our chances of success are determined by the logic of our approach to our problems. We must determine logically what actions must be taken, and then make sure that everyone involved knows what must be done to achieve our goals. In addition, we must make sure that we do not overlook some important function that could make the difference between success and failure. To organize the basic steps, that logic tells us must be accomplished if we are to be successful in fire prevention, we can make a function chart.

The basic functions of a fire prevention bureau are shown in Fig. 7.1. If the fire prevention bureau can carry out all these func-

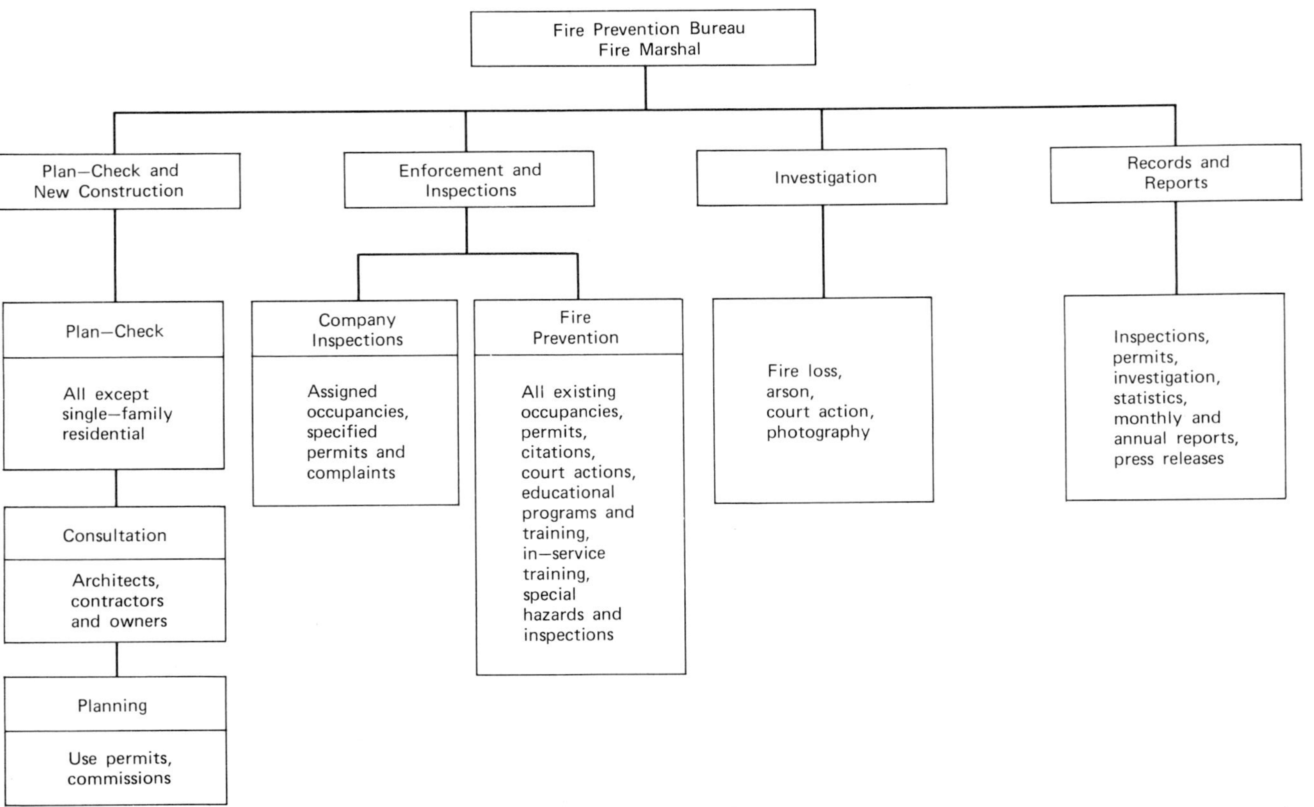

Fig. 7.1. The functions of a fire prevention bureau.

tions, then it probably has an effective program. Let's examine each of the functions listed in a bit more detail and determine their relationship to one another and to fire prevention objectives in general. We will begin with plan checking and new construction.

Plan Checking and New Construction

One of the most necessary and important functions of fire prevention is the checking of plans and new construction. In fact, if there is any hope for stemming the tide of fire loss and fire incidence, it is through this function, even if it means that less time is spent in other fire prevention functions. It is clear that unless buildings are constructed properly, with all the fire protection features inspected and installed properly, buildings soon become candidates for fire statistics. In addition, the buildings must be protected during construction, for they are not immune to fire caused by the carelessness of those working on its construction or by arson. In order to assure, as much as is humanly possible, that all the fire protection requirements are met in the construction of a building, the construction plans must be checked for conformance to both building and fire codes. Therefore, most fire prevention bureaus have a plan checking specialist; in a small bureau the fire marshal or one of his assistants usually checks construction plans. In a fire department that has a good working relationship with the building department, plans must be approved by the fire marshal before a permit to build is issued.

A spin-off of plan checking is that of providing information to the architects, engineers, and contractors during the design stages of a building regarding fire protection needs and requirements. This consultation saves, in the long run, money for the building owner and, in addition, establishes a rapport between the professional community and the fire department. Of course, in order to check plans properly and accurately, the members of the fire prevention bureau must have a good working knowledge of building construction, building and fire codes, and the ability to read and interpret construction plans.

Fire prevention personnel must also be familiar with the design and standards of built-in fire protection systems (e.g., automatic fire sprinklers, dry chemical systems, fire alarm systems), because they are required to check and approve installation plans for each of the fire protection systems required in the fire codes.

The process, if it is to be effective, cannot stop once plans of construction and fire protection systems have been checked. Field inspections must be made of new construction and installations and tests conducted; final acceptance must be based upon what actually is in the field, to insure that the approved plans have been met.

A conscientiously pursued program of checking plans and new construction, via field inspection, can help the fire department, and the fire prevention bureau in particular, do its part in building a safer community. A well-constructed building, with all the fire protection devices and systems installed and functioning properly, will reduce the local fire problem and reduce possible life loss in the event of fire. For example, the suppression crews, in the event of a fire, will not have to deal with inferior construction that aids fire spread. (More about building construction and fire spread is presented in later chapters.) From plan checking and new construction we move on to enforcement and inspections.

Enforcement and Inspections

These functions of a fire prevention bureau make up the bulk of the routine work and require manpower and a large number of man-hours. Enforcement and inspection are also the areas through which a company fire inspection program can have the greatest impact.

Enforcement and inspection include regular fire prevention inspection of all existing buildings, with the exception of single-family houses. (The chapter on "Codes and Ordinances" will explain more about what inspections are required.) In any community, with the exception of very, very small towns, the number of existing buildings (as opposed to new construction) to be inspected is almost overpowering for the manpower usually allotted to fire prevention. Therefore, given the proper training, the fire companies can aid in achieving a complete inspection program. However, the Bureau personnel cannot neglect special inspection needs that may be beyond the scope of fire company inspections; so the inspection workload may be reduced in some inspection areas but not totally eliminated.

Inspection or any fire prevention effort without enforcement is an exercise in futility. During inspection, violations of the fire code, as well as other codes, are noted and the building owner and/or occupant is given suggestions and/or corrections to comply with the codes. If fire prevention personnel do not follow up with reinspection to check compliance, then very soon the citizens of the community will realize that they really do not have to comply—"no one will check anyway." In the event that legal action is required (citation, court action), unless the fire department can show that every attempt was made, short of legal action, to obtain a correction and that reasonable time was granted for correction, two possibilities arise. First, if fire and loss of life occur, the fire department can be found liable for nonenforcement of the fire code and may even be cited for negligence if it has allowed any hazardous condition to exist with its knowledge and has not attempted to gain correction. Secondly, the building owner and/or occupant can claim that he was never notified (this shows the importance of written records and reports) or that he didn't know the violation had to be corrected by a certain time, as no one returned to find out about the correction. In this case, the fire department not only looks foolish, but endangers its chances in legal action against the code offender. Therefore, inspections and enforcement go hand in hand and form the basis for legal action directed at obtaining corrections and fire code compliance for the safety of all citizens.

Inspections and enforcement utilize the imagination and knowledge of the fire prevention officer to the extent that he must use his insights and familiarity with codes and fire to help the building owners and/or occupants engineer solutions to the fire problems found during inspection, and suggest viable alternatives that are both reasonable and economical to help assure fire safety. From this point, let's move on to records and reports.

Records and Reports

As in any organization, public or private, paper work flourishes. Some of the forms and reports are necessary, and some of it tends to be a bit of a waste because the information is never used. Nonetheless, the written word remains one of the cheapest yet most lasting forms of communication that we have available. In fire prevention, records and reports serve several functions, which we will outline in this section. We begin with the inspection report.

The inspection report is the backbone of fire prevention records and serves as the basis for several activities. First of all, the inspection report is written proof that an inspection was conducted and a report of what was found during inspection. In many respects it is similar to the prefire plan utilized by the fire suppression companies. A sample inspection report is shown in Fig. 7.2; it includes what was found and suggestion for corrections. It also indicates when a reinspection will be made to determine compliance, and is signed by both the inspector and the representative of the business being inspected. A copy is given to the building owner and/or occupant, and one is retained for the fire prevention files. In the event of legal action to gain correction, the copy or copies of the inspection and reinspection reports are utilized as proof that the inspection(s) was(were) made, and that due process was followed. The inspection report information is also used for statistical analysis of the kinds of violations found, the corrections obtained, and the number of inspections made monthly and annually, which are utilized in the fire prevention reports to the chief of the department. In addition, when the fire department undergoes its periodic grading by the Insurance Services Offices, these reports act as proof of an inspection program, and may be used to determine the efficiency of the fire prevention program as outlined in the Grading Schedule (see Chapter 3).

Other records and reports utilized by fire prevention include the following:

Building record file
Investigation report (fire)
Citation
Notice of violation
Complaint and action form

These are just samples of the types of records used; some de-

ISSUED BY ____________________ DATE ____-____-____

STATION # ________ RE-INSPECTION DATE ____-____-____

FIRE SAFETY NOTICE

BUSINESS ____________ ADDRESS ____________ OCCY. ____ CODE ____

CITY ____________ PHONE ____________ SIGNATURE ____________

______ **A. A REASONABLE DEGREE OF FIRE SAFETY EXISTS AT THIS TIME**
______ **B. SOUTH SAN FRANCISCO FIRE CODE REQUIRES COMPLIANCE WITH THOSE ITEMS INDICATED BELOW:**

1. ELECTRICAL:

______ a. Discontinue use of extension cords in lieu of permanent wiring . . . (27.4046)
______ b. Remove extension cords run through openings or attached in series . . . (27.4046)
______ c. Remove extension cords placed under rugs, furniture, or where subject to damage . . . (27.404a)
______ d. Maintain wiring in good condition and protect from damage . . . (27.404a)

2. EXITS:

______ a. Remove all other locks or latches from doors with panic hardware . . . (1.2019)
______ b. Remove obstructions from exits, aisles, corridors, and stairways . . . (10.103)
______ c. Unlock all exit doors during business hours . . . (10.104b)
______ d. Provide lighting for corridors, stairways, and exterior exitways . . . (10.113a)

3. FIRE ALARM SYSTEMS:

______ a. Maintain fire alarm system in operable condition and test monthly . . . (13.307c)

4. FIRE DOORS:

______ a. Remove obstruction(s) and alterations to fire doors . . . (10.104e)
______ b. Keep attic access and scuttle openings closed . . . (27.408)

5. FIRE EXTINGUISHERS:

______ a. Provide ______ extinguisher(s) of a ____________ minimum rating (____________ type) . . . (13.301a)
______ b. Mount extinguishers where readily available, with top not higher than 5 feet . . . (13.301a)
______ c. Post signs indicating location where extinguishers are not readily visible . . . (NFPA 10)
______ d. Service and tag (by State Licensee) each extinguisher annually and after use . . . (13.302)

6. FIRE PROTECTION INSTALLATIONS:

______ a. Maintain access to, and operation of standpipes, fire hose, and sprinkler control valves . . . (13.302)
______ b. Inspect and test sprinkler system monthly and maintain records . . . (13.302)
______ c. Identify sprinkler valves and secure in open position . . . (13.302)
______ d. Provide spare sprinklers (6 minimum) and sprinkler wrench . . . (13.302)
______ e. Inspect and service hood and duct extinguishing system over cooking equipment annually and after use . . . (13.302)
______ f. Remove decorations or other attachments from sprinkler system . . . (13.302)
______ g. Replace damaged, corroded or painted sprinkler heads . . . (13.302)

7. FLAMMABLE LIQUIDS:

______ a. Reduce Class I (gasoline, etc.) to less than 5 gal. inside & 10 gal. outside without a permit . . . (15.103)
______ b. Reduce Class II, III (solvents, etc.) to 25 gallons total inside & 60 gallons outside without a permit . . . (15.103)
______ c. Remove flammable liquids not stored in-proper containers . . . (15.104)
______ d. Remove flammable liquids not used for maintenance in assembly bldgs., offices, apartments and motels . . . (15.403b)
______ e. Discontinue use of Class I (gasoline, etc.) for cleaning . . . (15.111)
______ f. Store flammable liquids away from corridors, main aisles, stairways, and exit areas . . . (15.402a)
______ g. Discontinue dispensing from containers exceeding 5 gallons or provide pump taking suction from top . . . (15.108)
______ h. Discontinue the discharge of flammable or combustible liquids into drains or on the ground . . . (15.110)

8. HEAT PRODUCING APPLIANCES:

______ a. Remove combustibles and storage from heater area or room . . . (27.406d)
______ b. Provide clearance between heat producing appliances and combustible material . . . (27.406)

9. HOUSEKEEPING (INTERIOR):

______ a. Remove or store rubbish, waste material, and oily rags in closed metal containers . . . (27.201b)
______ b. Clean grease filters, and hood duct systems over cooking appliances . . . (27.407)
______ c. Provide waste collecting receptacles at dust producing machines . . . (27.201a)

10. RUBBISH (OUTSIDE ACCUMULATION):

______ a. Remove waste, trash, and weeds, OR store in closed metal or metal-lined containers . . . (27.201)

11. SMOKING AND OPEN FLAMES:

______ a. Post and enforce "NO SMOKING" signs ____________ . . . (29.101)
______ b. Provide ash trays, sand receptacles, or butt cans, where smoking is permitted . . . (29.102)

12. STORAGE:

______ a. Remove storage from exits, aisles, corridors and on or under stairways . . . (10.103)
______ b. Arrange storage in orderly manner and provide for exiting and Fire Department access . . . (1.210a)
______ c. Remove storage to 18" below level of sprinklers (36" for storage piled over 12 feet high) . . . (13.302)
______ d. Reduce storage height to at least 2 feet below ceiling . . . (27.203b)
______ e. Secure and identify compressed gas cylinders with name of product . . . (8.104 & 8.108)

13. PERMITS:

______ a. Obtain a permit from the Fire Prevention Bureau, or DISCONTINUE ____________ . . . (1.302)
______ b. Comply with permit addendum(s), # ____________, Item(s) ____________ . . . (1.306)

ADDITIONAL COMMENTS AND/OR REQUIREMENTS:

____ ____ ________________________________
____ ____ ________________________________
____ ____ ________________________________
____ ____ ________________________________

25. ☐ Letter will follow explaining additional requirements and establishing a re-inspection date.

RE-INSPECTION BY ________________________ DATE ____-____-____

VIOLATIONS CORRECTED ____________ NOT CORRECTED ____________ COMMENCE LEGAL PROCEDURE ☐

Fig. 7.2. An example of a "check-list" fire prevention inspection form. (Courtesy South San Francisco FD.)

partments use others (or the same records called by other names). Those that are able to use electronic data processing (computers) may have special forms. The importance of the efficient use of records and reports cannot be emphasized too strongly. Utilization of these records can determine specific information about the local fire problem and the efficiency of the fire prevention program, and therefore can be used as a basis for guiding the local fire prevention program to meet the community needs, as well as for those specific purposes mentioned previously (e.g., enforcement, legal action). From here we will move on to the public education function of fire prevention.

Public Education

Although public education is the responsibility of everyone in the fire department, the bulk of the work is often the responsibility of those assigned to fire prevention. Examples of public education programs conducted by those in fire prevention include:

School fire drills and fire safety education
Hospital fire drills and fire safety training
Private fire brigade training
Fire safety talks and demonstrations to service and other community organizations

In many fire departments the training officer and fire suppression personnel augment fire prevention personnel in providing fire safety demonstrations, such as extinguisher-use training and private fire brigade training. The basic purpose behind public education is, of course, to provide information to the public regarding fire prevention and safety. It has been shown that if people know about problems and are shown what to do, they will, for the most part, take steps to prevent problems from occurring. In other words, if the community is aware of the need for having escape plans to evacuate their homes in the event of fire (Operation EDITH—evacuation drills in the home), and they are educated to know what to do in the event of fire, there will be less chance of fires occurring or people being trapped in their homes in the event of fire.

Another benefit of education programs is that they keep the fire department in the public eye. This is especially true if the fire department has a good working relationship with the news media, and its programs are covered by the local news. People will be aware of what is being done and also feel that their fire department isn't just sitting around in the firehouse waiting for a fire. If the public feels good about their fire department, then pay raises and other benefits are easier to obtain, and a good rapport is established between the fire department and the citizens. Of course, in order to gain news media attention, it helps to give out periodic news releases, and in many instances the writing of news releases is part of the job of fire prevention. Releases from fire prevention are especially useful during National Fire Prevention Week. In preparing news releases and providng other public education programs, the records and reports can be utilized for specific information that pertains to the local fire problem. Our last section on fire prevention is devoted to fire investigation and complaints and weed control.

Fire Investigation and Weed Control

The reason for combining these last two functions is that weed control and complaints are in many instances handled as part of inspection and enforcement. We want to note here that fire investigation is sometimes a function of an arson squad or bureau, but in many instances it is still the responsibility of those in fire prevention. Fire investigation will be covered in more detail in the next chapter.

Weed control and complaints are often one and the same thing. Weed control is an attempt to reduce the number of grass fires on vacant lots in a city. In some cities this is handled by the public works department, but in others the fire department is responsible. In essence, the operation of the weed control program entails having lots cleaned, either by hand or by a tractor, to turn the weeds under before they dry and become a potential fire hazard. In the city that I worked for we had a contractor who did the work for us, and the cost of the work (for those who didn't clean their own land) was put on the tax assessment rolls and paid for by the owners when they paid their property taxes. Many weed control programs create fire breaks on large segments of land, such as pasturage. In our city it did reduce grass fire calls to some extent, but it was a time-consuming job for the bureau and required paperwork and supervision of the contractor. However, I must say that it did clean up the city's vacant lots. A second kind of complaint really goes hand in hand with public nuisance control.

The department receives calls or notices from the public about anything a citizen feels is a fire hazard. The calls may refer to neighborhood feuds, piles of waste deposited on lots and in yards, and so forth. The complaints are primarily from residential areas where people feel there is a hazard to their property and themselves. Usually, a complaint form is filled out; then an inspector is sent out to determine if there is a problem and what to do about it. Usually tact and diplomacy on the part of the inspector are required to resolve a neighborhood problem. Many times the problem is settled using the same inspection and enforcement process as previously outlined, but some situations require a peacemaker. Since we are talking about some of the roles played by a fire prevention officer or inspector, it is appropriate to discuss here the personal requirements for fire prevention.

PROFILE OF A FIRE PREVENTION OFFICER

As you have probably gathered from our discussion of the functions of a fire prevention bureau, anyone working in fire prevention must be a person of many talents. A fire prevention officer must be knowledgeable about fire behavior and hazards, must be able to speak and write effectively, must be conversant with building codes, fire codes, and national standards for fire protection, must be knowledgeable about law enforcement and the requirements for legal action, and must have a working knowledge of fire investigation. In essence, he must be a true professional to do an effective job in fire prevention. In addition, in many communities the fire prevention officer is the public's most constant contact with the fire department; therefore, his appearance and actions will be judged by the public as representative of the fire department. Inside the department, he must communicate effectively with his coworkers and in many cases provide the training in fire prevention presented to the fire companies (i.e., suppression crews).

A fire prevention officer usually works a 40-hour week, with additional duties, such as fire investigation that may increase that number of hours. His pay as well as his rank vary from department to department but usually are above that of top hoseman or fire fighter and may be equivalent to a captain. The head of the bureau is usually a chief officer—except in small departments—he works nominally a 40-hour week.

As mentioned in the beginning of the chapter, entry into the fire prevention bureau is usually via a civil service promotional examination. Depending upon the department, men may be assigned for specific periods of time in fire prevention without taking a promotional test. I believe that all future fire department officers should spend at least a year to two full time in fire prevention to enable them to become well-rounded professional officers and eventually chiefs of departments.

Fire prevention was the first area in the Fire Service to open up to a trained woman. In fact, I take pleasure and pride in the knowledge that one of my fire science students was one of the first, if not the first, woman to be hired by a fire department as a full fire prevention officer (Karen Straub of the Stanford University Fire Department, Palo Alto, California. Since she was first hired, she has moved to the Santa Clara County Fire Marshal's Office, and from reports she is fulfilling her duties as well as any man).

It is encouraging to note that the past practice of putting men into fire prevention because they can no longer perform fire fighting duties, because of injury, is beginning to recede. Most of these men didn't want to be in fire prevention in the first place, and you can imagine the impression they gave the public. Those who I have spoken to resented being forced into the job because they couldn't obtain a disability allowance, and they were somewhat bitter. It is far better to have an interested person in the job, rather than just a body to take up space and be an asset on paper.

Now, before we close this chapter, let's briefly discuss a company fire inspection program (fire suppression forces utilized for fire inspection work).

COMPANY FIRE INSPECTION

The involvement of fire companies in inspection work is growing annually; this a result, in large measure, of the growing demand for the efficient use of manpower for dollars spent by both the taxpayer and the city management. In addition, increased emphasis on fire prevention is one of the needs identified in the President's Commission report "America Burning." Given that the need for fire company inspection is valuable and necessary, let's examine a company inspection program organization.

It is logical to have the coordination of a company fire inspection program guided by the fire prevention bureau. The use of fire prevention records can help assure efficient scheduling and control of the inspections performed by companies. Problems and legal action that may be beyond the capabilities of the companies can then be handled by those in fire prevention. The day-to-day responsibilities are best handled by battalion chiefs within their respective districts, and by the captains and lieutenants of the engine companies. As an example of a company fire inspection program, I'll use the one that I was involved in while a Fire Marshal.

In this particular program the city was divided into inspection districts that were essentially the same as the first-due areas of responsibility of each station. Thus, the engine companies were not out of their primary response area when inspecting, and in the event of a call they could respond directly from their inspection location without significant time loss. The primary inspection districts were then further divided into smaller areas that were the responsibility of the company officer on each shift; as there were three shifts, the primary areas was divided into thirds.

The fire prevention bureau then published a monthly list of buildings (i.e., addresses, business names) that were due for inspection and a list of reinspections that were due, based on prior monthly inspection reports. Any stubborn noncompliance problems or unusual problems were referred to the fire prevention bureau by the company officers (in this case captains).

The captain of each shift was responsible for the inspection scheduling, and the completion of inspection reports and the recording of other data. The captains scheduled the inspections (a minimum number per month) to fit into his crew's schedule of training and other duties. The other members of each crew (fire fighters, drivers) were assigned specific duties such as drawing or correcting a prefire plan, noting violations, or whatever tasks the captain deemed appropriate. During an inspection, the driver was left with the engine, and he and the captain were in constant radio contact with the fire alarm center. (The captain had a hand-held radio.) If an alarm came in during the inspection, it was transmitted over the radio, and the engine company on inspection could respond from the location of inspection if required. As far as I can recall, emergency calls seldom interrupted inspections, which was due to either good scheduling by the captains or the effectiveness of the prevention program. To avoid undue agitation of passersby, banners were placed on the engines during inspection tours. It also provided good fire department advertising at low cost.

So, as can be seen from this example, everyone had a responsibility and a job to perform during inspections, and everyone was involved with the prevention program. With good planning and training a fire prevention program can be enhanced by a total fire department team effort, and the response from citizens is usually very good for the fire department. Such a total prevention program can also improve the fire department grading classification (see chapter 3), which could benefit the whole community by a reduction in fire insurance rates. In addition, a full inspection program can reduce the number of fire calls in a community, as well as provide a significant reduction in fire loss. Isn't that what fire protection is all about?

SUMMARY

Fire prevention is the first line of defense of any fire department. How well the fire prevention program works is determined in part by the chief of the fire department and his goals. However, except in very small fire departments, the fire marshal (fire prevention officer) is the one who is actually responsible for the day-by-day activities of the bureau.

Regardless of the size of the fire department, the functions and objectives of the fire prevention program are the same. In essence, the purpose of a fire prevention program (and a fire prevention bureau) is that of saving lives and property from fire. This is achieved by performing the functions of plan checking, inspection, issuing permits, enforcement, and public education. To achieve an effective prevention program, good written records are a must. These records and reports, combined with fire investigation, help identify the local fire problem, and, if utilized properly, can aid in the reduction of fire loss in the community and serve as a basis for public education and special inspection programs.

As with any effort, the total involvement of all fire department personnel can increase the effectiveness of the fire prevention program. Through the use of a company fire inspection program guided by the Fire Prevention Bureau, a well-rounded and productive fire prevention program that meets the communitys needs can be achieved. In addition, there are spin-off benefits from a total program, such as increased fire department productivity and improved community relations. It is far better and safer to stop a fire with pen and paper than with nozzle, water, and hose. No one has died because a written notice required the clearing of a blocked exit, but people have died when they have not been able to get through a blocked exit during a fire.

REVIEW QUESTIONS

1. Where does the legal authority for the position of fire marshal (head of the fire prevention bureau) come from? Who has the ultimate responsibility for a fire prevention program?
2. What is the purpose of the fire prevention bureau? How does it differ from the overall purpose of a fire department?
3. List the functions of a fire prevention bureau and briefly describe each one.
4. Who is responsible for the inspection of existing buildings?

5. What are the basic needs for an effective enforcement program? Why?
6. What methods can be used to pinpoint the local fire problem? How are they used?
7. Outline the basic duties and qualifications for a fire prevention officer. What kind of person is needed for fire prevention?
8. Define in your own words a company fire inspection program.
9. Briefly outline the functions and operations of a company fire inspection program.

REFERENCES

William K. Bare, *Fundamentals of Fire Prevention,* John Wiley & Sons, New York, 1977.

International City Manager's Association, *Municipal Fire Administration,* Chapter 11, Chicago, IL, 1967.

National Fire Protection Association, *Fire Protection Handbook,* Sections 9 and 18, Boston. MA, 1976.

chapter 8

8 FIRE INVESTIGATION

Regardless of the number and excellence of fire fighters, equipment, and apparatus, and of the fire prevention programs that a fire department may have, fires still occur. Furthermore, when a fire department extinguishes a fire, the job isn't finished,—the cause(s) of the fire must still be investigated and the amount of damage must be determined. The department and/or other investigative agencies (such as the police department or state fire marshal's office) must determine if the fire was accidental or deliberate. If the fire was accidental, then the cause is important to help prevent similar fires. If deliberate, the cause is important to aid in establishing the crime of arson.

In this chapter we will look at the investigation of fire in relation to the total fire protection system, the purpose of investigation, general investigation procedures, and who is responsible for investigation. We will also discuss the general qualifications required for a fire investigator.

At the end of this chapter, you will be able to:

1. Identify who is responsible for fire investigation.
2. Define a fire cause.
3. Identify the basic steps in the investigative procedure of a fire department.
4. Define arson.

5. Identify the roles of fire department members in fire investigation.
6. Identify the qualifications required to be a fire investigator.

New Terms

Fire cause	*Investigation report*
Point of origin	*Arson*

WHO IS RESPONSIBLE?

In a general way, everyone working in a fire protection agency, from the head of the department to the newest fire fighter, is responsible for some aspect of fire investigation. In actual practice, the brunt of the responsibility for the investigation is usually assigned to a specific person or persons, in a fire department, a local police agency or to members of both the police and fire departments. They then form an arson or investigation squad.

No matter who is assigned the responsibility, complete fire investigation requires training and total fire department involvement. To better understand this point, let's identify how the actions of a fire department may affect an investigation; at the same time, we will outline the steps in an investigation.

THE INVESTIGATION PROCESS

We know that in order for a fire to start we must have an ignition source and a fuel source to begin the ignition sequence. Investigation consists of a series of steps taken to determine the fire cause (ignition source plus fuel source), point(s) of origin, and who, if anyone, may have been responsible for the fire.

The first step in the investigation process begins at the time the alarm is received by the fire department. If the alarm is phoned in, the alarm operator records the vital data pertaining to the call such as the kind of fire (e.g., structure, grass, automotive), the location, the time the alarm was received, the date, and, if possible, the identity of the person who reported the fire. In many departments all calls on the emergency phone number are tape recorded, making the alarm operator's task a bit easier. If the calls are not tape recorded, then the operator must make a written record of the information in the alarm log.

The next step is observation; that is all fire department members must observe what they see, hear, and smell, from the time they leave the station until they return. The following are some examples of what fire fighters and company officers should observe when going to and while at the fire scene:

- Any cars that may follow the apparatus to the scene.
- The color of the smoke on arrival, location of the visible flames, amount of fire spread, and any obstructions encountered at the scene, such as cars obstructing access to driveways.
- Actions taken upon arrival by each individual fire fighter and company (i.e., all operations such as forcible entry into buildings, location of furnishings, and the base of the fire).
- Anyone who seems overly anxious to help or volunteer information, or who has been seen at other fires recently.

All of these observations may become important in the investigation of a fire and should be remembered. Obviously it is a matter of training; a fire fighter cannot take time to write down notes on his observations while fighting a fire, but with proper training he can observe much and record what he has noticed later.

During fire fighting operations the investigation process continues; fire fighters and company officers take care not to disturb a fire scene any more than is necessary, in order to preserve the scene intact. In addition, after all the actual extinguishing operations have been completed, the company officers and fire fighters not engaged in other tasks begin the actual on-site investigation and try to determine the fire cause. If the officer in charge needs help, then the fire investigator (usually a chief officer, fire marshal, or someone from fire prevention or the local police agency) is called to the scene; when he arrives, he assumes charge of the investigation.

During the on-site investigation, everything at the fire scene is examined, and to the trained investigator, the story of the fire unfolds. It may take a few hours or it may take weeks to determine the cause, but the investigator examines the flame-spread pattern, depth of charring, and other factors until the point of origin is found. Everything, including statements of witnesses and actions of fire department personnel, is documented by photographs and by written record and/or tape recordings. At the origin, all possible accidental causes are examined until the correct one is found. If no possible accidental causes are found, then the fire may have been caused deliberately; that is, arson may be indicated.

Arson Investigation

If arson is suspected, then the investigation becomes both a fire and a criminal investigation. In most states, arson is defined by the penal code, and the penalties for arson are set out. There may be degrees of arson, with different penalties for each degree, but the basic definition of arson is usually the same: "the deliberate, willful burning of property or thing that belongs to another." Within this definition the burning of a "thing" includes buildings, automobiles or other equipment, forests, rangelands, food crops, and structures such as bridges, fences, and railroad trestles. If a death occurs as a result of arson, a murder charge is also a possibility.

The investigation process remains the same for routine fire investigation and arson investigation up to the point of determining whether the fire was caused by arson. At that point, arson investigation becomes the same as criminal investigation. Suspects are found and evidence is gathered; for a conviction it must be proven beyond a reasonable doubt that there was intent to commit arson, that damage resulted from the fire, and that the suspect(s) did start the fire. This section of arson investigation is more police than fire department work, and for that reason many fire departments may ask for police aid, or turn all the fire investigation findings over to the arson squad or police detectives for follow-up and prosecution. As mentioned above, even if arson is not suspected, the findings of a fire investigation are important to a fire department and its operations.

Investigation Records and Their Uses

The findings of fire investigations are important not only to a fire department but to others interested in fire protection. Insurance companies utilize investigation findings to help determine enforcement and settlement of fire insurance policy provisions. Investigation reports are used by a fire department as an aid in determining the local fire problem and as a guide for an effective fire prevention program. If the causes of local fires are identified, this knowledge can make fire inspections and public education more relevant, and can help reduce the incidence of particular fire causes.

Fire investigation reports, as implied above, are also used in legal actions. The recorded investigation findings are important in both civil (insurance suits) and criminal (arson) actions.

Another use for investigation findings is statistical analysis at the state and national levels. For example, the National Fire Protection Association couldn't give a statistical analysis of the national fire problem each year without the findings of state and local investigations. However, for this information to be useful, the reports must be factual and accurate, and based on sound fire investigation practices used by well-trained fire departments and fire investigators.

REQUIREMENTS FOR FIRE INVESTIGATION

A good fire investigator will have the following qualifications and training:

- Knowledge of fire behavior, fire causes, and fire suppression operations. (This is usually by a combination of schooling and experience in actual fire fighting.)
- Knowledge of and training in fire investigation methods.

- Knowledge of and training in criminal investigation methods, including preservation of evidence, laws of arrest, and other criminal processes.
- Familiarity with court proceedings, legal requirements, and case management.

As is apparent by now, if one hopes to be a fire investigator, at least some time will have to be spent in a fire department before one can hope to step into a fire investigator's role. However, one may also choose police training first and then specialize in fire investigation.

SUMMARY

Although an assigned fire investigator may have the primary responsibility for fire department investigations, every member of the department must be aware of investigation needs and procedures. In order to be able to find the cause of the fire and point(s) of origin, the investigation of a fire must begin as soon as possible after the alarm is received.

Fire department members must be trained to observe conditions, during response and while at the fire scene, to aid investigators. In addition, fire fighters must take care during fire fighting operations to insure, as far as possible, preservation of evidence. When the fire is out, examination of the scene coupled with reports from witnesses can be used to pinpoint what happened and why it happened. If no accidental causes can be found, then the investigation shifts to searching for possible arson.

An arson investigation combines the principles and methods of fire investigation and criminal investigation. The definitions and penalties for arson are set out in the state penal code and indicate how a case must be built for prosecution. Because arson is a crime, cooperation of the local police agency should be welcomed. If the fire department has an investigator trained in both fire and criminal investigation, so much the better.

The findings of all investigation reports (both accidental and arson) are important for pinpointing the local, state, and national fire problems; these reports are also necessary for any legal action that may be started.

A fire investigator must be well versed in all aspects of fire protection, as well as a well-trained investigator. He must also be familiar with legal proceedings and case management.

Investigation is an important aspect of the total fire protection picture, and it should not be handled lightly or carelessly.

REVIEW QUESTIONS

1. Who usually has the primary responsibility for fire department investigations?
2. What are the purposes of a fire investigation?
3. Outline the steps in an investigation and explain the importance of each step.
4. As a fire fighter, what should you be aware of when responding to an alarm?
5. At what point in an investigation should arson be suspected? Why?
6. What constitutes a case of arson?
7. Does an arson investigation differ from a routine fire investigation? If so, explain the difference(s)?
8. Why are investigation records important? How can they be used?
9. List the essential requirements for being an investigator and your reasons why the requirements are important.

REFERENCES

Paul I. Kirk, *Fire Investigation*, John Wiley & Sons, New York, NY, 1969.

William K. Bare, *Fundamentals of Fire Prevention*, John Wiley & Sons, New York, NY, 1977.

International City Managers Association, *Municipal Fire Administration*, 7th ed., Chapter 12, Chicago, IL, 1967.

National Fire Protection Association, *Fire Protection Handbook*, 14 ed., Section 1, Chapter 14, Boston, MA, 1976.

chapter 9

9 FIRE DEPARTMENT MANAGEMENT

The guiding force that controls all the subsystems involved in a fire department is the administration and management—that is, the chief of the department and other fire department administrative personnel. In addition, since no system stands alone, there are other outside influences that affect the management of a fire department, such as city and state political bodies, labor unions, and laws.

In this chapter we will examine the functions of fire department management and how they are or can be implemented. We will also discuss the responsibilities of the chief, such as budgeting, interdepartmental relationships, and manning. We will also look at some of those outside forces that determine or influence fire department management.

The objectives of this chapter are:

1. To identify the basic functions of management and their relation to the objectives and structure of a fire department.
2. To identify the role and functions of the chief of the department in relation to intra and interdepartmental concerns affecting fire protection.

3. To identify some of the outside influences that affect fire department management.

New Terms

Functions of management	*Planning*
Organizing	*Motivating and directing*
Controlling	*Management cycle*

MANAGEMENT FUNCTIONS

A fire department must adhere to certain basic principles of good management in order to provide the best fire protection possible, through the best use of manpower, equipment, and money. These basic management principles—which can also be applied to our private lives—are:

Planning
Organizing
Motivating and Directing
Controlling

Let's examine each of these activities in relation to a fire department.

Planning

Part of planning is defining what you are trying to accomplish and then deciding how to carry through your plans to reach your goals. Planning also includes listing the steps that must be taken to reach the goals. Some examples of the steps needed when planning for a fire department are:

- Developing short- and long-range goals and plans.
- Preparing budgets which include the purchase of new equipment and apparatus.

- Defining the purpose and goals of each position in the fire department in relation to the objectives of the fire department.

In well-managed fire departments, everyone works to define and plan goals, and each person knows his part and importance to the fulfillment of the departments goals.

The next function we discuss is organizing.

Organizing

The manner people and materials (resources) are used to carry out plans is the main concern of organizing. Organizing helps define the authority and responsibility of each position and how best to choose personnel to fill these positions. Organizing also helps in

- Establishing training programs
- Establishing communication and a good working relationship with other departments within the city (building, planning, etc.), and with other fire protection agencies
- Gaining supplies and funds to carry out plans
- Establishing a maintenance program for buildings and equipment
- Establishing standard operating procedures for response and action at the scene of an emergency

Now that we discussed planning and organizing our resources, our next step is to describe motivating and directing.

Motivating and Directing

This is "where the action is" in the management cycle. Motivating and directing are the functions of management that get things done through directing, coordinating, and motivating personnel to achieve the department's goals and to fulfill their individual responsibilities.

The ways these functions of management are accomplished include:

- Defining the goals of the total department in terms of company and fire prevention objectives.
- Fire prevention code enforcement, investigations, and inspections
- Company prefire planning and evaluation of company emergency operations and procedures
- Supervising emergency operations
- Counseling of personnel
- Manning of apparatus and setting up work hours

These activities must be carried on every day, every month, every year, but the only way a chief of a department or any individual can know if these functions are effective is by checking periodically; in other words, by controlling.

Controlling

Controlling amounts to evaluating performance and gathering information to improve results and attain the overall objectives. Some of the ways controlling is accomplished are:

- Establishing rules, regulations, and standard operating procedures for performance during regular duties and at the scene of emergencies
- Evaluating records and reports
- Evaluating performance reports on personnel
- Controlling budget expenditures
- Revising plans to meet new or revised objectives.

The results and information received through controlling are then used by the chief as a basis for further planning, and the cycle of management continues. Management does not follow a set course, nor is it a process that can be used to maintain the status quo. It is a dynamic process, and one that can yield greater fire protection, greater productivity, and greater job satisfaction for all members of the department.

Now that we know about the basic functions of management, let's look at some examples of what a chief and his administrative staff must contend with in order to manage a fire department.

ADMINISTRATIVE FUNCTIONS

One of the first functions that comes to mind is dealing with intradepartment personnel. A chief of a department must deal with manpower problems and labor relations. Almost every department has a fire fighters' association or a union, or both. The employee groups have a decided effect on management, because the union or association brings to the chief grievances to be resolved, annual requests for salary increases and fringe benefits, input as to the type of manning and work week that it would like utilized by the fire department, and other labor problems.

Most individual employee problems are handled through the chain of command, but may be, as a last resort, taken directly to the chief of the department. However, it is usually standard procedure for the chief to review all dispositions of employee problems. Furthermore, the chief must review annual employee performance reports and consider and approve promotions and pay increases. Of course, any reduction in work hours or change in wages or fringe benefits must have an effect on his budget planning; therefore an annual review of employee's desires and needs is essential.

It is in the area of budgeting that the chief usually has his toughest time. A fire department rarely produces revenues (unless they charge for permits), so city fathers usually require a great deal of proof of need and of productivity before any budget increase is allowed. Since most departments draw from the general fund, the competition for funds is keen. A wise chief will use his administrative and field officers to help him prepare and compose budget justifications, which not only help the chief but the whole department as well.

Funds are also needed for capital improvements. Capital improvements include items that have a long term use such as new apparatus, additional or replacement equipment, and building of new or repairs of existing fire stations. Because of the cost of these items, a chief is required to make and recommend long-range plans that include setting aside a specific amount per year toward these improvements. For example, a new engine costs between $60,000 and $80,000. In order to purchase a piece of apparatus at that price several thousand dollars per year must be set aside. The chief must know when he will need the engine, must write up the specifications for the apparatus, put the specifications out for bids, then select the manufacturer that can meet both the specifications and the dollar limits. Other factors that a chief of a department must consider when planning for capital improvements are the growth of the jurisdiction, other changes that may require new stations and apparatus, and estimated costs for future fire protection. Needless to say, in order to sell his program a chief must be able to communicate his needs.

One of the most important functions a chief performs is communicating with the local political body, the citizens, and with his personnel. Quite a bit of the chief's time is spent communicating with others in meetings and via the media as well as through written reports. In essence, an effective chief is a combination of a good fire fighter, politician, and manager. A chief must also spend time attending council meetings, staff meetings, and public appearances. Since the chief usually serves at the pleasure of an administrative official of the local political body, many times his ability to keep his job is dependant upon keeping the official (city manager, mayor, chairman of the fire district, town council) informed for the fire department's achievements, productivity, needs, and goals. One of the ways this is done is through written reports.

Usually, a chief must submit, at the minimum, monthly reports and an annual report, in addition to budget justification. In all three instances the chief must rely on reports from his administrative staff, and these are based on the company run reports, fire investigation reports, fire prevention reports, and other data collected by the department. The chief must also utilize fire department reports when called upon to speak on an issue affecting fire protection. An example of an issue that affects fire protection is that of the rezoning of property within a city. The rezoning may increase the fire protection needs for a particular portion of the city or district; therefore, the chief must be able to tell the planning commission and other political bodies how the planned changes affect fire protection. Another issue is the requirement of automatic fire sprinklers for particular types and uses of buildings. The chief must be able to justify the need for a change; if he fails, the requirement would be dropped.

As is probably apparent by now, other governmental departments and jurisdictions affect fire protection, and because of this, the chief must make sure that all agencies (e.g., building department, police department, public works, planning) have a close working relationship with the fire department; this prevents conflicts.

It is part of a chief's job to make certain that everyone is informed of fire protection activities and requirements. The media must be used to inform the public of fire department activities, fire prevention tips, fires in progress, to name a few topics.

At this point, let's remember that the chief is also in command at fires, and fire suppression duties must still be fulfilled. However, except in very small fire departments and districts, chiefs of departments should not have to take charge of every fire. As elsewhere, on the fireground good management must be the rule. After all, if the chief is a good manager, his officers and men should be able to handle the fire without his directing all their actions in every instance. This is not to say that the chief can avoid directing very large fire operations. In large fires, he is usually found at a readily available command post, directing overall operations from that location. If the chief finds himself on the end of a hose line fighting a fire, he is in big trouble. He had better look at his organization again to see what went wrong or, better yet, ask himself if he should really be there. I worked under a chief who couldn't keep away from the nozzle, and who had to take charge of every fire. His impact on the department and myself was one of resentment and ill will. It killed the initiative on the part of everyone and morale suffered as well as performance. This instance brings us to the subject of administrative assistance. A chief of a department cannot be effective if he attempts to be a one-man band—he needs help.

Administrative Assistance

To carry out management functions and to accomplish all the duties of a chief requires that others in the department be utilized in various administrative capacities. We have noted in previous chapters that the assistant chief, fire marshal, and others help carry the load of running a department. In addition to the help of these people, a chief will also have the aid of a secretary and possibly an administrative assistant or two. These administrative assistants may be fire department personnel on special duty or they can be civilian employees. The chief may also receive help from the personnel department and other administrative agencies within the jurisdiction.

Further assistance, without increased manpower, can be gained from the utilization of electronic data processing (computers), where available. By putting records and reports on a computer program, a chief can have an analysis of fire department operations very quickly and clearly. The computer can reduce expenditures for personnel who would be required to manually scan and compile all the information that can be contained in a readout of a good program.

SUMMARY

Effective fire department management requires that the chief of the department and his officers understand and utilize sound management practices. A fire department, just like any other business, must have a budget and handle employees efficiently so as to gain maximum productivity.

In order to achieve efficiency, productivity, and effectiveness, there must be a plan, good organization, and motivation for employees to achieve department goals and objectives. In addition, a system for checking on progress is required to prevent stagnation and to redefine goals when required. A department and its management should be dynamic, rather than dedicated to maintaining the status quo. In instances where management stultifies employee participation, morale problems increase and rapid turnover of employees becomes a problem. Neither effect (low morale and turnover) benefits the citizens or the department.

A chief of a department is required to perform many

administrative tasks. Included in his job are dealings with labor and government management (e.g., city manager, town council, fire commission), budgeting, capital improvement planning, and communication with the public at large. The chief is also required to submit various types of reports on fire department activities, which are then used by the city manager as some measure of fire department productivity. In addition, the chief is responsible for fireground activities, including fire fighting, fire investigation, and all fire prevention activities.

Quite obviously, one man cannot do it all; therefore, the selection of good officers and staff is necessary for good management and administration. Other sources of administrative assistance for a chief include skilled secretarial help and, where possible, use of electronic data processing.

REVIEW QUESTIONS

1. What are the basic functions of management?
2. Explain how management functions are used to establish department goals and objectives.
3. List the administrative responsibilities of a chief of a department. Explain how each one affects management.
4. What are some items that would be considered fire department capital improvements?
5. What are the items used for a chief's report(s) of fire department activities?
6. Explain why intradepartmental relationships are important to a fire department.
7. What are the sources of administrative assistance for a chief of a department?

REFERENCES

International City Managers Association, *Municipal Fire Administration,* 7th ed., Chapters 3 and 7, Chicago, IL, 1967.

National Fire Protection Association, *Fire Protection Handbook,* Section 9, Chapter 2, Boston, MA, 1976.

PART III

III FIRE PROTECTION AND THE ENVIRONMENT

chapter 10

10 CODES AND ORDINANCES

This chapter begins our survey of the environmental factors that must be dealt with in fire protection. One of the keystones for controlling the hazards of the environment are the laws and regulations that apply to building construction and contents. Further, there are also laws regulating exterior (outside of buildings) and wildland hazards that are enforced by a fire protection agency. Although other regulations control working conditions, retirement,labor contracts for fire fighters, our focus here will be confined to building and fire regulations.

The objectives of this chapter are:

1. To define a code.
2. To identify the different kinds of codes that affect fire protection.
3. To identify general contents of building and fire codes, and their purpose.
4. To identify how codes and nationally recognized standards work together.
5. To identify who is responsible for code enforcement.
6. To identify the relationship between state and local regulations.

New Terms

Code	*Code enforcment process*
Building code	*Field inspection*
Fire prevention code	*Show-cause hearing*
Standard	*Citation*
Municipal code	

WHAT IS A CODE?

Before we begin our exploration of codes, we should define what a code is. In essence, a code is simply all the laws and regulations (minimum requirements) that pertain to a specific subject, placed in book form with an explanatory title. For example, a building code deals with building construction; within its covers you will find all the regulations pertaining to construction, as well as references to standards and other codes which supplement the building code provisions. The same applies to all codes, such as fire prevention and fire protection, electrical, plumbing, and municipal codes.

So far so good, but we have to go a step further. First, there is no universally adopted building or fire code, and second, each level of government has its own set of codes and minimum standards. Let's examine each of these statements in a bit more detail.

TYPES OF CODES

In the United States there are several different codes within a specific subject. For example, we find the following building codes, each with its own publisher and groups of followers:

Uniform Building Code	International Conference of Building Officials
National Building Code	American Insurance Association
Basic Building Code	Building Officials Conference of America

Southern Standard Building Code	Southern Building Code Conference

There are fewer fire codes, but no nationally accepted one. The major fire codes are:

Uniform Fire Code	Western Fire Chiefs's Association and International Conference of Building Officials
Fire Prevention Code	American Insurance Association
National Fire Codes	National Fire Protection Association
BOCA Basic Fire Prevention Code	Building Officials Conference of America

Furthermore, jurisdictions may adopt a code and then amend it to suit their particular needs, which leads to further loss of standardization in codes.

The next set of problems we consider occurs at state and local government levels.

State and Local Codes

Each level of government, including the federal government, attempts to regulate the actions of its citizens. The federal government sets certain minimum standards for building construction and fire protection and leaves other regulatory requirements to the individual states. The states are given the power (by the Constitution of the United States) to adopt laws and enforce them for the health, safety, and welfare of their citizens. Thus, each state has its own minimum standards for statewide regulation of building construction, fire protection, vehicles, and other aspects of our lives. Most states have a state fire marshal's office to insure that minimum state fire safety and protection standards are written and enforced. Thus, uniformity throughout the state is achieved, unless local political bodies enact more restrictive regulations.

Where state codes are silent, or more restrictive local requirements are desired, the state allows local political bodies (county and city governments) to adopt local codes and regulations. The local requirements and codes must be equal to or more restrictive than the state minimum standards. In other words, a local code cannot be adopted which allows conditions to exist that are forbidden by the state adopted minimum requirements.

Local Codes

The basic code for local regulations is the municipal code or its equal. The content of the codes vary, but it will contain all the basic regulations of a city or other local political jurisdiction. These regulations include the ordinances that signify the adoption of local codes (e.g., building, fire, electrical, plumbing codes), sanitation regulations, zoning ordinances, civil service policies, and other similar types of regulations. Thus, if you want to know what building code or fire code the jurisdiction has adopted,you can look in the municipal code or its equal and find the specific ordinance which states the code, the edition, what amendments were made (if any), and any special enforcement procedures. Of course, a simpler way would be to ask the people who are charged with the enforcement of the specific code.

At this point we are ready to look at the content of building and fire codes which may be adopted locally.

CODE CONTENTS

Each Code adopted locally will contain specific regulations on particular subjects (e.g., fire prevention), indicate who is responsible for enforcement, outline record keeping requirements, and provide references to nationally recognized standards that are used to supplement the code. Since most codes follow essentially the same format, let's examine a typical building code and a typical fire code.

Building Code

The responsibility for the enforcement of the building code and other construction codes is normally that of the chief building inspector and his staff. However, when a good working relationship exists between the fire and building departments, consultation between the fire marshal and the building official can benefit the community. Most building codes do provide for joint enforcement and consultation; however, it can only be implemented if cooperation is mandated by the city government.

The administrative section of a building code gives information on joint enforcement, what permits are to be issued, the fee schedule for permits, inspection and enforcement methods, and penalties for violations of code provisions. (Most building and fire code violations are misdemeanors and are punishable by revocation of permits, by fines, and by jail terms of up to six months.)

Within the body of the building code, we find the basic construction requirements; they are based on the location of the building in relation to the property line, use of the building (occupancy classification), type of construction (fire-resistive, combustible), materials used, number of stories, and total floor area. There are also requirements for exiting, earthquake protection, automatic fire sprinkler systems, and other life safety and property protection items.

A building code defines the different types of occupancies and types of construction, and states what other codes and standards must be met. For example, the plumbing must conform to the local plumbing code, the electrical system to the local electrical code (usually the National Electrical Code, published by the National Fire Protection Association). Conformity to the proper code is required for the heating and ventilating systems, automatic fire sprinkler systems, and other systems. Standards, such as those established by nationally recognized testing agencies (e.g., Underwriters Laboratories, Inc.) for construction materials, are used as a yardstick for conformance with code requirements. Fire codes are similar in many respects, except that they concentrate more on fire prevention and protection requirements.

Fire Codes and Fire prevention

The chief of the department and his representatives (e.g., fire marshal, inspectors, fire suppression personnel) are responsible for the enforcement of the fire code. In the administrative section of a fire code we find, in essence, the outline of what a fire department is supposed to accomplish. Some of the duties required of a chief are suppressing fires, prevention of fires, control of hazardous conditions, elimination of fire hazards (both outside and inside of buildings), investigation of fires, and the keeping of records and reports on all activities. In addition, a chief and his representatives are given the right to enter property to make inspections, provided that the inspections are made during reasonable hours and the fire department has reason to believe that hazards exist within or on private property.

As in the building code, permits are required, and specific regulations for achieving greater fire safety cover such topics as :

Flammable liquids
Explosives
Fire protection equipment
Fumigation
Hazardous chemicals
Cryogenics
Places of public assembly

For the most part, the fire code deals with the contents of buildings and the surrounding properties and leaves the actual building construction requirements to the building codes and the building officials. The fire code can be applied to existing as well as new conditions and can therefore be termed retroactive in nature. In essence, then, a fire code is a maintenance code, because it pertains to maintaining fire sfety at all times, in structures with all types occupancies,regardless of the age of the building.

The fire code, along with the building code, states the standards to be used as references for further specific requirements and determination of code conformance.

Since we have learned that standards supplement code provisions in both the building and fire codes, let's examine an example of how this works.

Codes and Standards

Codes often lay out general requirements, but leave the specific minimum requirements to a referred standard. If this wasn't done, a code would be a huge document, and perhaps be more confusing than it already seems to most people not directly involved with code enforcement. To illustrate the relationship of a code and a reference standard we will use an example from a building code.

In the building code, there is a section that establishes when and where automatic fire sprinklers are required. The codes tell the reader that for given types of building uses, and with certain construction conditions, automatic fire sprinklers are required; however, the code does not say how the system shall be constructed. Instead it refers to a standard contained in the National Fire Codes.

This standard, NFPA No. 13—"Sprinkler Systems, Installation," contains the specific requirements for constructing and installing sprinkler systems for different conditions. It states what plans are required, who must approve the system installation, the testing required, the spacing of the sprinkler heads, the size of the pipe and water supply, and other information. Therefore, if a sprinkler system is installed as directed by NFPA No. 13, it is deemed as meeting that particular construction code requirement.

As is obvious, codes and standards are interwoven to form a total pattern of minimum fire and life safety, as well as construction standards that are for the benefit of everyone. However, all the written words in the world will accomplish nothing unless someone checks to make sure the codes are being met—this brings us to code enforcement.

CODE ENFORCEMENT

Codes can be made effective only through enforcement. The backbone of any enforcement program, regardless of the code involved, is a group of well-trained, knowledgeable people who are responsible for inspection, permit issuing, plan checking, and other constrols used to achieve code conformance. Let's examine an example of the typical code enforcement process as used in the field for fire prevention.

First, let's assume that we have had a request from a business owner to store a large quantity of gasoline at his place of business. The fire code requires that a permit be obtained from the fire department before he can take any action related to gasoline storage. Therefore, he comes to us and applies for his permit; he gives us plans for the installation and fills out an application.

Someone is assigned to check the plans and the application. A field inspection is also made of the storage site to determine if basic code requirements for gasoline storage can be met and what changes must be made or conditions corrected before the permit is issued. When the permit is issued and the storage facilities are completed, another inspection is made before actual storage can begin; if there are no problems the final permit approval is given. Periodically thereafter, someone from the fire department will make an inspection to assure that the code requirements are still being met and that any hazards found are corrected.

Let's assume that at some time the installation is found to have certain violations; the owner is advised by the inspector what conditions must be corrected and how long before a reinspection will be made. The owner is also given advice on how to accomplish the corrections so he can have some idea of what can be done. If the corrections are made, the cycle of periodic inspections continues. If the corrections are not made within a reasonable time, then we enter another phase of the enforcement process.

At this point several consequences may occur. They are:

1. His permit may be revoked and he may be directed to cease operation.
2. He may receive a citation to go directly to a court of law.
3. He may be required to show cause why his permit should not be revoked and his operation allowed to continue. All of this is called a show-cause hearing, and is usually conducted before the city attorney and/or county counsel.
4. He may have a hearing, and appeal to the chief of the department to continue his operation and keep his permit.

One or more of these steps may be taken by the fire prevention bureau, as directed by the policy of the particular fire department.

The main point is that the code must be enforced; if it is not, no one will pay any attention to safety requirements.

In our next chapter we will examine building construction, thereby adding to our knowledge of codes and how they affect the fire protection problem and our environment.

SUMMARY

All of our laws, regardless of concern, are found within the covers of varous codes and under titles that group together almost all the laws pertaining to specific subjects. In addition, codes provide references to other related codes and standards to ease code enforcement.

Codes relating to fire prevention and protection are found nationwide and at all levels of government. We can divide these codes into codes pertaining to building construction and codes pertaining to fire prevention and protection. Several codes of each type, published by different sources, are available for adoption by governmental agencies; therefore, nationwide code standardization has not been realized. Furthermore, legislatures at federal, state, and local levels enact additional laws for fire protection and building construction, which adds to the maze of code requirements.

In most states, an office of the state fire marshal handles the minimum statewide requirements for fire protection and prevention. However, each local jurisdiction has the option of adopting codes and ordinances, providing these codes are not less restrictive than state standards. Where state laws are silent, local jurisdictions have the option of setting out minimum local fire safety and other requirements. To find out which codes are adopted locally and what other ordinances pertain to fire protection, one can check the local municipal code or its equivalent.

Both building and fire codes contain specific information relating to the responsibilty for enforcement, the records and permits required, minimum requirements for construction, installation and use of buildings and equipment, and penalties for code violations. Codes also refer to other codes and standards that are used to supplement the original code provisions. Codes and standards form the basis for public safety, but they are useless unless they are enforced.

The code enforcement process may vary slightly from jurisdiction to jurisdiction, but the basic process remains the same everywhere. The process includes issuing permits where required, field inspections, reinspections, and if violations persist, legal action either through citation or show-cause hearings. The main point is that through enforcement, minimum fire protection and life safety can be attained and maintained.

REVIEW QUESTIONS

1. What is a code?
2. In your opinion, should there be only one standard code relating to fire prevention and protection available nationwide? Why? Why not?
3. Outline the division of regulatory powers beginning with the federal government.
4. Can a local governmental agency adopt a less restrictive requirement than the state requirement? Why?
5. List some common kinds of codes that are adopted at the local level.
6. In what section of a code do we find out who is responsible for enforcement? List some of the other items found in the same section.
7. Outline some of the items covered by a building code.
8. Outline the general content of a fire code.
9. Explain how codes and standards work together.
10. Outline the steps used in the code enforcement process.

REFERENCES

American Iron and Steel Institute, *Fire Protection Through Modern Building Codes*, 4th ed., New York, NY, 1971.

William K. Bare, *Fundamentals of Fire Prevention*, Chapter 4, 9, and 10, John Wiley & Sons, New York, NY, 1977.

National Fire Protection Association, Fire Protection Handbook, 14th ed., Section 6, Chapter 11, and Appendix D, Boston, MA, 1976.

International City Managers' Association, Municipal Fire Administration, 7th ed., Chapters 10 and 11, Chicago, IL, 1967.

chapter 11

11 BUILDING CONSTRUCTION AND HAZARDS

The knowledge of basic building construction features and of content hazards in general, is necessary for fire prevention and fire suppression. This knowledge is also necessary for code enforcement, and it provides a key to determining how a fire may spread in any given building, which will affect a fire department's tactics and strategy in the event of fire.

A building or structure is only as good as its construction and maintenance. Therefore, a building may aid in fire spread or, if constructed and maintained properly, it can confine a fire to a small area. Both the construction and contents affect life safety for both the occupents and the fire fighters who must extinguish a fire.

In this chapter we will:

1. Identify the effects of construction on fire spread.
2. Identify types of construction.
3. Identify the kinds and importance of occupancy classifications.
4. Identify the major components of construction and how they affect the fire problem.
5. Identify the purpose and major steps in prefire planning.

New Terms

Occupancy or type of use	*Heavy timber*
Type of construction	*Unprotected combustible*
Noncombustible	*Rated*
Protected combustible	*Nonrated*
Fire-resistive rating	*Interior finish*
Flame-spread rating	*Prefire plan*
Target hazard	*Marking system*

THE ENVIRONMENT

As we look around us, we see an amazing array of man-made structures. The most common structures are the buildings, whose design and use vary greatly. For example, compare the building in Fig. 11.1 with your home or apartment and with the building that you are in now; each fulfills its special requirements for its particular use.

To go a step further, Fig. 11.2 shows the Transamerica building control room, which regulates heating, ventilation, security, and fire alarm and extinguishing systems, among other functions. Compare this with the simple controls for the heating and ventilating systems within your home. Quite a difference! Each is designed and constructed to meet specific requirements, yet they all have one thing in common—danger from fire. Each building must be constructed so that in the event of fire the rapid spread of fire will be prevented and the occupants will have sufficient time to reach safety. In addition, it must contain the fire, if possible, until a fire department can arrive and extinguish the flames. How is this accomplished? What are the factors that must be considered? How do they affect fire protection?

To answer the above questions, let's examine the major factors in building construction, their impact on how a building is built, and how they assist or retard fire and affect life safety.

Fig. 11.1. The "Pyramid" building—San Francisco offices of the Transamerica Corporation. (Courtesy of the Transamerica Corp.)

Fig. 11.2. The control room of the "Pyramid" building which handles all the various systems of the building including the security and fire protection systems. (Courtesy Transamerica Corp.)

OCCUPANCY

The term occupancy refers to the type of use of a building or structure. For example, it relates to whether the building is required to seat thousands or hundreds of people (public assembly), to store hazardous materials (flammable liquids, explosives), or to contain many items for sale, plus storage (retail sales store). Each type of occupancy mentioned presents both life and property safety problems. However, in a building used for public assembly, life safety is the main consideration, whereas in the storage of hazardous materials, life safety remains a problem, but the main focus of concern is on the rapid and destructive spread of fire due to the nature of a building's contents. Looking at our third example, retail sales stores, we must consider both life safety and rapid fire spread, but again the contents concern us more, from a fire standpoint. Walk into any supermarket and you can find a great quantity of items for sale, including potential explosives, e.g.,spray cans, and poisons, e.g.,pesticides and cleaning agents. On the other hand, the total number of people inside at any one time is considerably less than you would find at a theater; therefore, contents become our primary consideration.

Poisons and explosives in a supermarket? That's right—and those items become dangerous during a fire. If you are on the end of a hose line you'd better be thinking about contents. Docile spray cans, subjected to heat, explode and become flying metal, and they may even become miniature projectiles that can cause serious injury. Vaporized cleaning agents and houseplant pesticides will be in the atmosphere you breathe. The type of occupancy is important from a code enforcement standpoint, and is of the utmost importance to fire fighters' safety.

For code enforcement officials, insurance underwriters, and building contractors, the projected use of a structure is one of the keystones that determine the type of construction required and influence the design of the building. For those involved in fire suppression, it is a factor that will determine the tactics and safety precautions to be used in the event of fire. How a fire is fought is influenced by construction, and that is the subject of our next section.

TYPE OF CONSTRUCTION

Type of construction refers to all those components of a building or structure which form the skeleton of a building. In describing type of construction we must consider the roof, floors, walls, doors, windows, stairs, and elevator shafts. These are the main components of the building that will either withstand and confine a fire or will allow a fire to spread and destroy the entire building and its contents, including those people who live or work inside. These structural elements form a chain of protective components which will be as strong as its weakest link. Fire will find the weakest link in any building. With this in mind, let's examine the common types of construction used today.

Construction is classified by its capacity to resist fire. The designations used for specific classification vary; however, general categories can be listed as follows:

Fire-resistive
Noncombustible or rated
Protected combustible or rated
Heavy timber
Unprotected combustible or nonrated

Each general classification tells us a bit about what we can expect from the building in the event of fire. Either the building will withstand a fire and contribute little or no fuel to the fire or it will fuel the fire and may be subject to total collapse and total destruction during a fire. The building codes refine these general categories, so let's discuss them a bit further.

Fire-Resistive

Fire-resistive construction is also known as Type I and Type II construction. Structures of these types, if built and maintained properly, confine a fire and do not add any fuel (in the form of combustible structural components), as structures of the other types may. In Fig. 11.3 we see a typical building of reinforced concrete, classified as Type I or fire-resistive (F.R.). Other Type I alternatives to concrete are protected steel and masonry. In Fig.11.4 we see two types of columns used in fire-resistive construction in a bit more detail; we are shown here an example of how the three-hour rating for the frame can be met. Type II fire resistive construction differs from Type I mainly in the lower fire resistive rating required; the construction materials remain essentially the same. (Fire-resistive ratings will be discussed in greater detail later.)

Noncombustible

Type II construction, both one-hour rated and nonrated, fit within this category. The typical frame, with structural components of steel or equally noncombustible material, is shown in 11.5. In Fig. 11.5*a* we see how the frame may be protected to achieve a rated construction, or left unprotected and classified as nonrated, Type II.

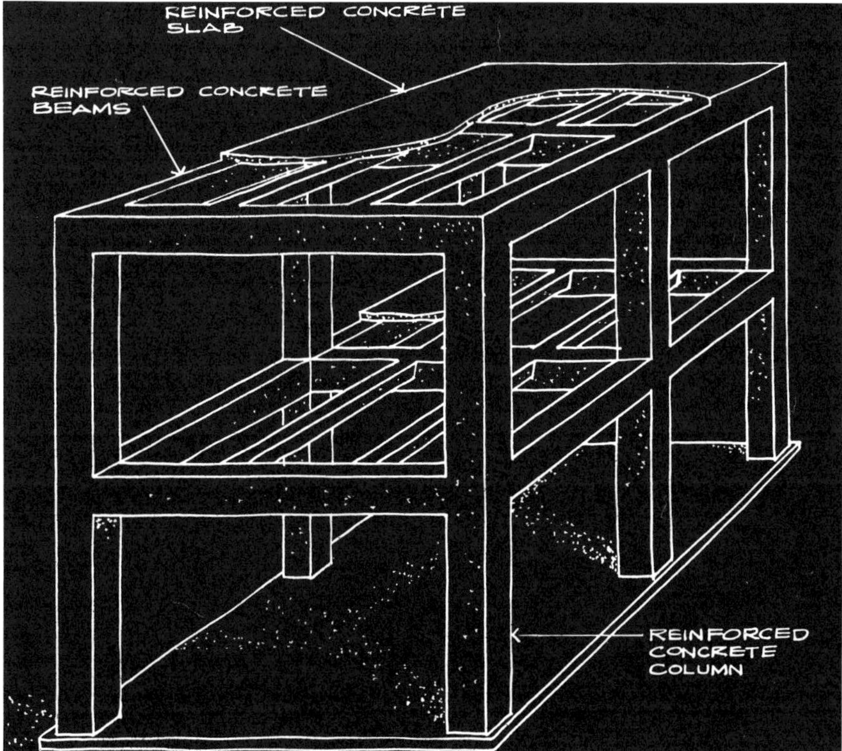

Fig. 11.3. Typical Type I construction. (Courtesy Spencer Marquis, Architect.)

Protected Combustible or Rated

Within this category are Type III, one-hour, Type III nonrated, and Type IV heavy timber construction. All three are then within the scope of protected combustible. In Fig. 11.6 the typical dimensions of lumber that are suitable for heavy timber construction are shown. Wood has a tendency to char, and thus protect itself from fire with the insulation provided by the char; this makes construction with heavy timber structurally sound, as compared with unprotected steel which can and does collapse in the event of fire. The exterior walls in Type III and IV construction are still required to have a minimum of a four-hour rating, which is usually gained via the use of masonry or concrete.

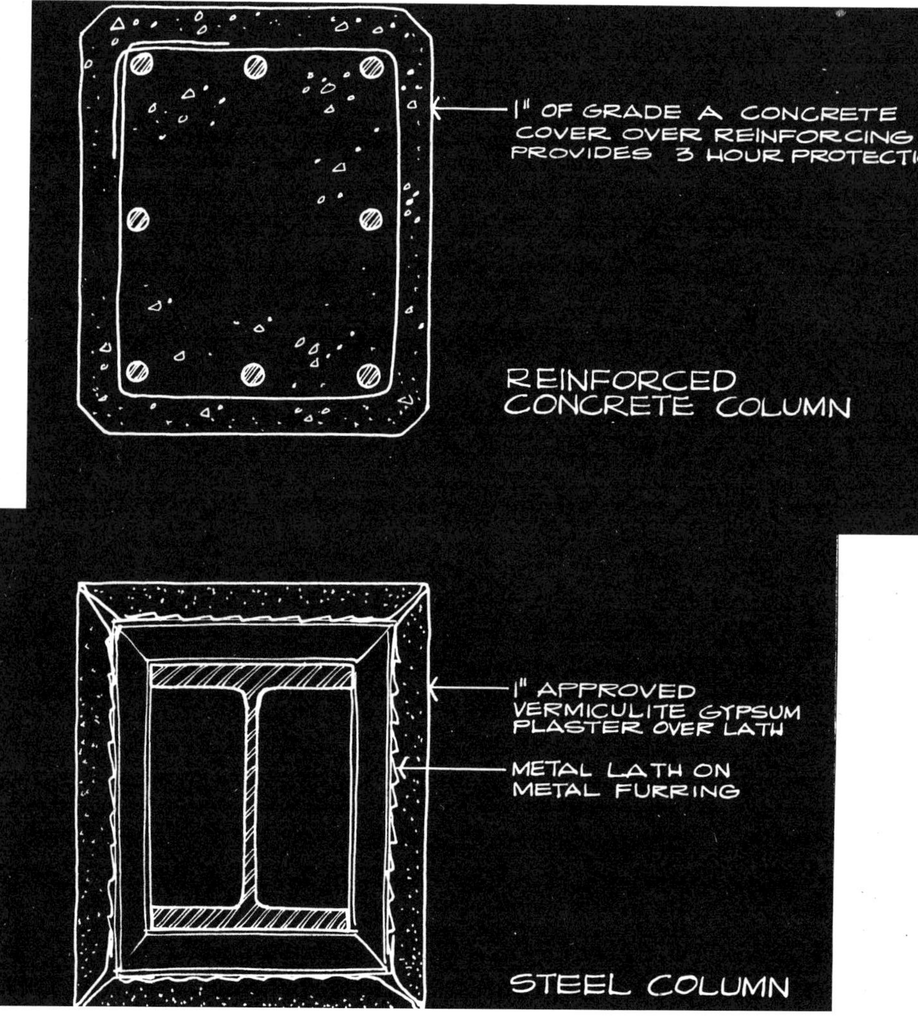

Fig. 11.4. Typical columns for a Type I construction. (Courtesy Spencer Marquis, Architect.)

Unprotected Combustible

Type V construction, the most common type of construction because it is used extensively in building private homes, is the last type of construction to be discussed. It may be either rated or nonrated (Fig. 11.7); the typical wood frame is shown in Fig. 11.7.

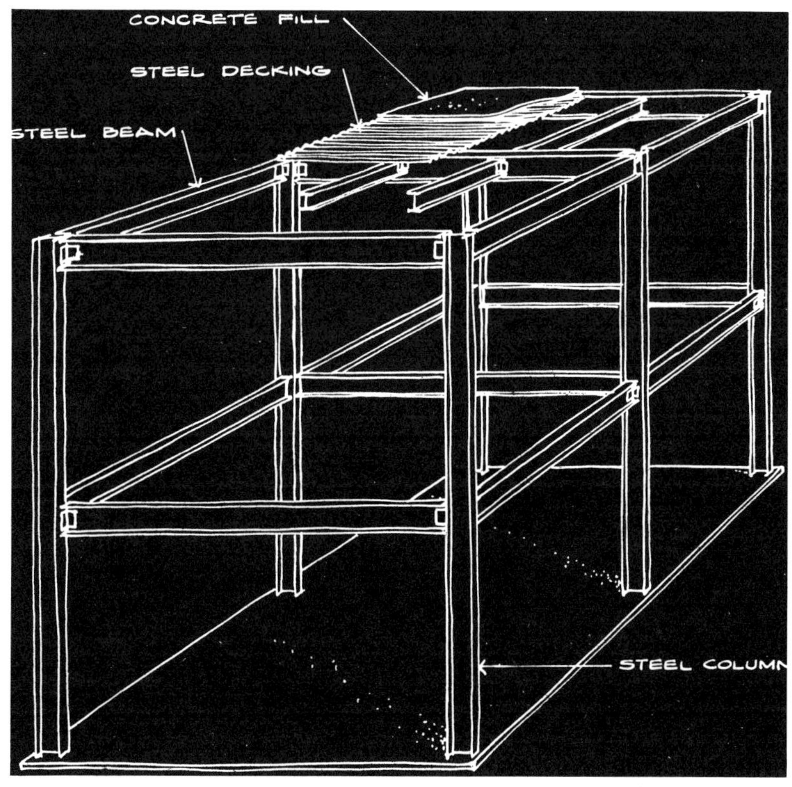

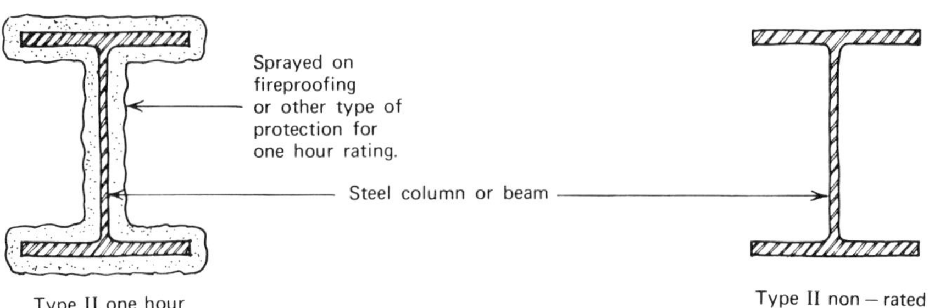

Fig. 11.5. *a* and *b*. Typical Type II construction. (Courtesy Spencer Marquis, Architect.)

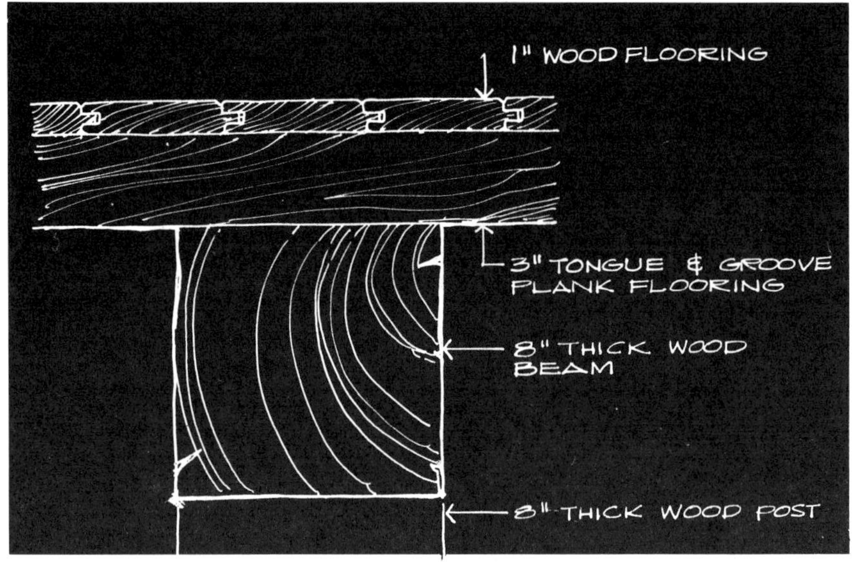

Fig. 11.6. Typical Type IV construction, heavy timber. (Courtesy Spencer Marquis, Architect.)

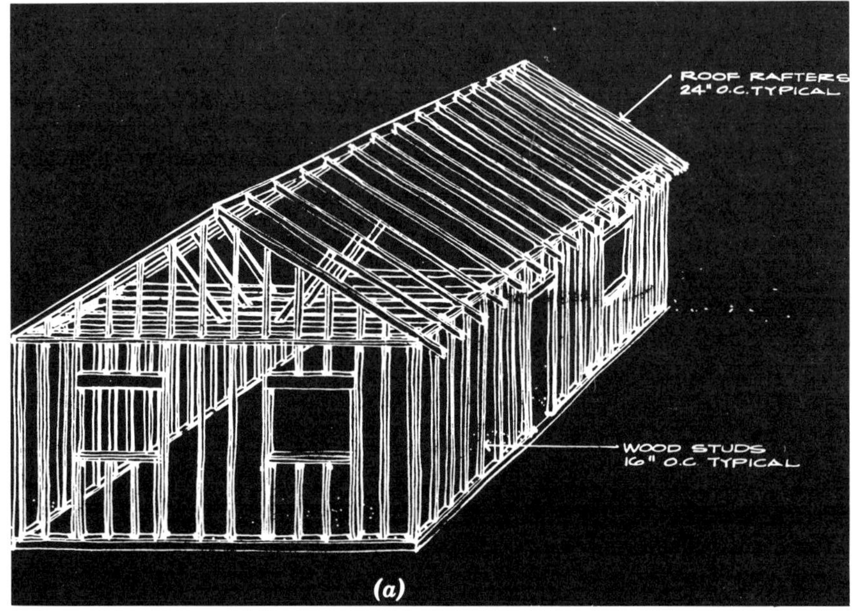

Fig. 11.7*a* and *b*. Typical Type V construction. (Courtesy Spencer Marquis, Architect.)

Unfortunately, both the contents and the construction members themselves add fuel to a fire; therefore, in terms of potential damage to the total structure, this type of construction rates high in fire risk. Its chief advantage is low cost (relatively speaking) to both the builder and the prospective home owner.

Now that we have covered the essential differences between types of construction, and we know that type of occupancy influences building construction, we see that these two factors com-

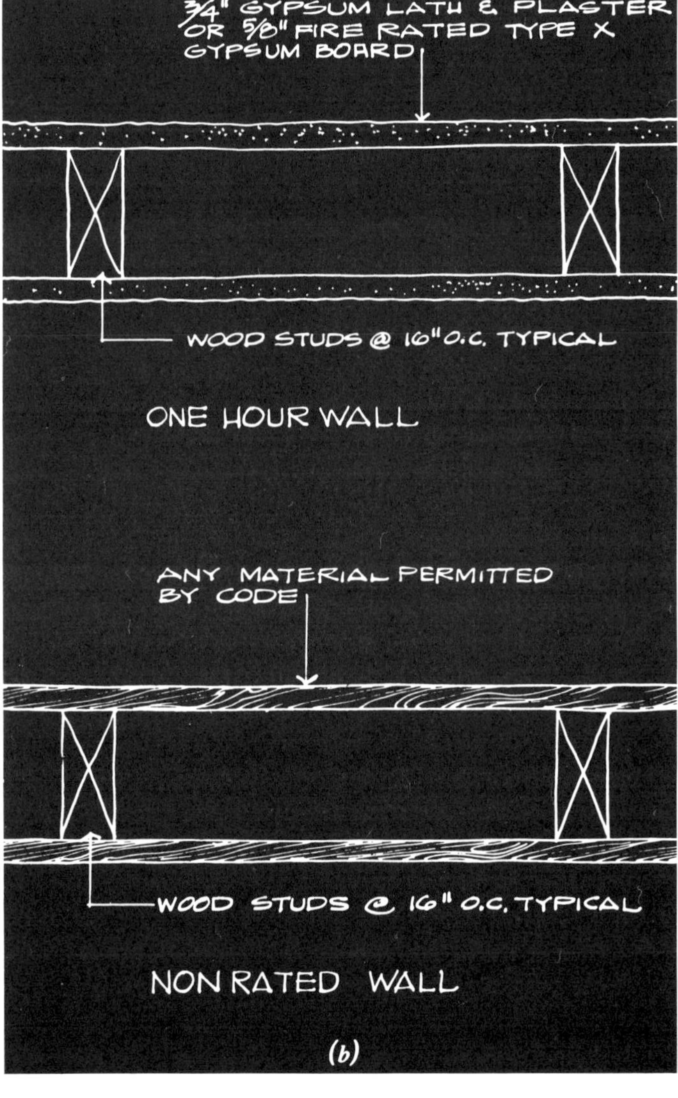

(b)

bined (type of construction and type of occupancy) determine to a great extent the allowable floor area and number of stories in a building.

Our discussion now turns to the ability of construction to withstand fire.

FIRE-RESISTIVE RATINGS

Our discussion of types of construction included the terms "fire-resistive," "rated," and "non-rated"; each of these terms relate to the ability (or inability) of a construction feature or component to withstand a fire for a given number of hours or fractions thereof.

In construction, walls, ceilings, floors, windows, doors, and other components such as fire dampers (located in heating and ventilating ducts)are fire resistive rated or nonrated. If a construction component must meet a specific rating, the building code will state the rating required in hours. For example, Type I fire-resistive construction must have the following minimum ratings for each component listed:

Component	*Rating, in hours*
Exterior walls	4
Frame (structural)	3
Shaft enclosures (stairs, elevators, laundry chutes)	2
Roof	2

To achieve these ratings, testing laboratories test walls and other construction components and designs submitted to them by manufacturers of construction materials. The components undergo an actual fire test to determine how long they will prevent fire from penetrating from one side of the component (e.g., wall) to the other. The amount of time this penetration takes is given as the fire-resistive rating (e.g., 2 hours means 2-hour fire-resistive rating). Then if a contractor or code official wants to know how to construct a 2-hour rated wall, he can refer to the testing agency listings, which show several approved materials and designs that carry the 2-hour rating. To be specific, Underwriters Laboratories, in its building construction listing, gives many types of approved rated walls and indicates exactly how they must be constructed and what specific materials must be utilized. Windows and frames, doors and frames, and other similar items are also tested and listed with their ratings expressed in hours.

In instances where the building code does not require a specific rating for construction, the builder has the option to choose any materials he may desire, as long as they are approved for use by the building code. Thus every aspect of construction is controlled, making possible both fire safety and structural stability.

As you have gathered by now, materials and designs with the fire-resistive rating are important in preventing or minimizing large fire loss; they help to confine a fire within the building of origin, provided the building is built and maintained properly. From a fire prevention standpoint, having rated construction aids in minimizing the hazards of building contents; from a fire fighting point of view these same rated items allow the fire fighters to gain the upper hand by helping to keep the fire in a limited area until extinguishing activities can begin. However, fire protection cannot rely solely on fire-resistive rated construction.

We now turn to other components in building construction that can affect the rapidity of fire spread: interior finish and contents. After discussing these, we will study a bit about exiting, that is egress and ingress for both the occupants and fire fighters.

INTERIOR FINISH, CONTENTS, AND EXITING

The interior finish (e.g., wall coverings, paint, curtains) is closely controlled in many types of uses but is strictly controlled in public assembly occupancies (e.g., theaters, auditoriums, restaurants). Building and fire codes provide minimum flame-spread ratings, for exitways and for decorative materials, that must be met for code compliance. These minimum flame-spread ratings, as determined by testing laboratories, are intended to reduce fire spread as much as possible. However, as buildings are used the original decorative materials and interior finishes may be changed, and as they are changed they may alter the rapidity of interior flame spread. Therefore, every effort should be made by those involved in inspection work to insist on maintaining the minimum required flame-spread rating on all decorative materials and interior finishes.

Along with the interior finish and decorative material, the contents provide fuels which may increase the flame spread within a building and add to the potential severity of any fire that may occur. For example, if storage of stock is haphazard and general housekeeping is poor, the physical arrangement and readily available fuel can help a fire advance rapidly. Furthermore, if the contents are subjected to readily available ignition sources, such as open flames, heating appliances and/or oxidizing agents, fire is almost assured of a rapid start and quick spread. If appliances and machinery are not properly maintained, if doors to stairs are not kept closed, if fire-resistive components are not maintained, the building cannot fulfill one of its functions which is containment of fire. Life safety will be threatened if the exitways are not kept clear of storage and stock and if doors to stairs and exit corridors are not tight-fitting, rated, and kept closed. If smoke, heat, and flame are allowed to penetrate exits, then panic, injury, and, possibly, loss of life may ensue. In addition, fire fighters may find their access to the building blocked because of the same exit obstructions that prevent the occupants from escaping. The only way these conditions can be held to a minimum, if not totally corrected, is through constant inspection by the fire department. An advantage of combining prefire planning and prevention inspections is that the fire suppression crews will be familiar with the individual buildings' contents and specific problems.

PREFIRE PLANNING

Prefire planning is just that—planning for a fire before it happens. The responsibility of prefire planning is usually given to each company officer and it is his responsibility to develop prefire plans, along with his crew, for the main buildings within his first-in response area. (These prefire plans are kept on file in each station, to allow company officers to become familiar with the work that has been done by others.) At the top of the list for prefire planning are those buildings that present unusual life or content hazards. These are commonly called "target" hazard occupancies. Target hazards include schools, theaters, bulk storage flammable liquid and gas plants, warehouses with highly valued stock, and manufacturing plants using flammable, toxic, or explosive materials. Other types of occupancies are also prefire planned, but are lower on the priority list than the target hazards.

To make a prefire plan requires an on-site building survey and/or inspection that is similar to a fire prevention inspection. Both include checks on the fire and life safety hazards and fire protection systems (automatic fire sprinklers, standpipes); the visits also provide opportunities to become familiar with the building and its contents. Both types of inspections study the building in a "what if" condition; that is, what if a fire starts? How can the fire be limited to prevent large loss and widespread damage? The main difference is that a fire prevention inspection seeks to correct hazards and improve fire and life safety conditions, whereas a prefire plan survey may be construed as an effort to become familiar with conditions to facilitate planning what to do if a fire starts, without seeking to correct the conditions found. However, prefire planning may correct conditions if it is the policy of the fire department to combine both prefire planning and fire prevention inspections. Be that as it may, let's examine prefire planning a bit further by discussing the basic steps utilized in the prefire planning process.

Steps for Prefire Planning

As stated above, the company officer and his crew must go to a building before a prefire plan can be made, and they must obtain permission to enter the building, once on the grounds. After gaining permission, they survey the building by going through all the areas, making sketches and taking notes on both the interior and the exterior features of the building. The following are some of the main observations that must be made for a complete prefire plan:

- Location of nearest hydrants, and best access for the apparatus.

- Location and types of private fire protection systems (e.g., automatic fire sprinklers).
- Location of power shutoffs (i.e., gas, oil, and electrical).
- Location, types, and amounts of hazardous materials (e.g., radioactive materials, toxic and corrosive materials).
- Assessment of the potential life safety problems and determination of the exit locations for possible entry and rescue needs.
- Location of any structural hazards that may become traps for fire fighters, such as dead-end corridors, high-piled stock, open pits, and elevator shafts.

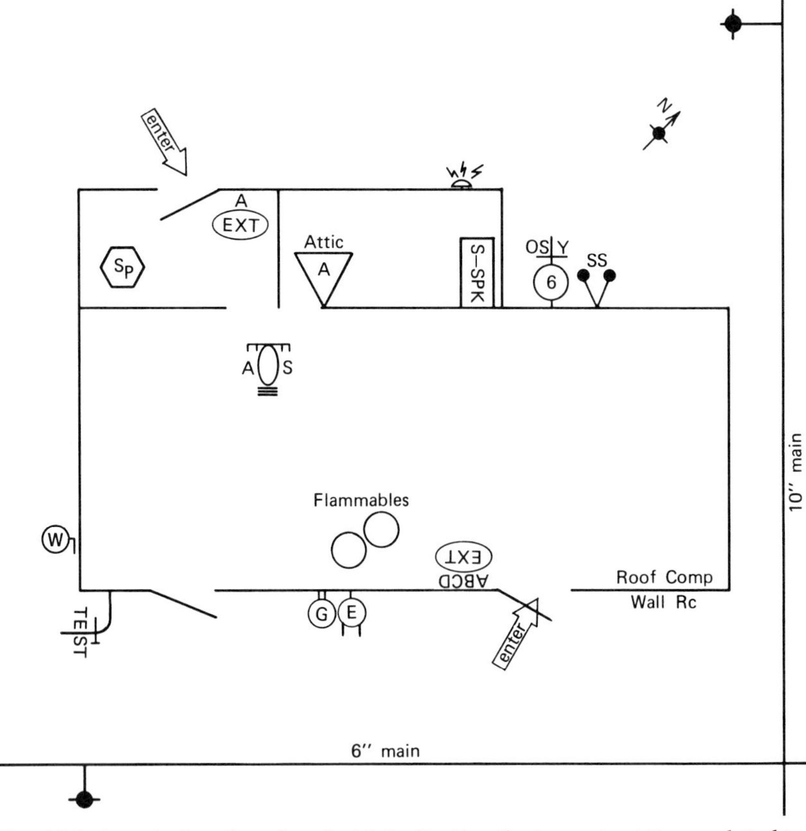

Fig. 11.8. A typical prefire plan sketch indicating the important items related to fire protection. (Courtesy Ron Coleman, Phenix Technology, San Clemente, CA.)

- Assessment of the danger to adjacent buildings or other structures from fire spread.
- Determination of type of construction, location of fire doors, and location of any obstructions that may interfere with fire fighting operations within the building.
- Location and number of roof vents and skylights, and assessment of any particular problems that might be encountered when ventilating (removing the heat and smoke) the building in the event of fire.

The above are only a few of the items to be noted and listed in a prefire plan report. In addition to the written report, a sketch or survey map is also made, by the fire fighting crews, of the property being prefire planned. The sample prefire plan sketch in Fig.11.8 notes the main factors related to fire protection, such as

KEY

(PLAN–PAK, Phenix Technology)

Symbol	Meaning	Symbol	Meaning
A EXT	Class A extinguisher	OS Y	Outside stem and yoke valve
enter	Where to force entry	S–SPK	Spare sprinklers
Sp	Salvage problem	A	Point of access
A S	Automatic sprinklers		Local alarm
W	Water shutoff	N	Direction of north
TEST	Alarm test valve		Three–way hydrant
	Two–way hydrant		
G	Gas shutoff		
E	Electrical shutoff		
ABCD EXT	All–purpose extinguisher		
Comp	Composition (roof)		
Rc	Reinforced concrete		
SS	Sprinkler connection		

hydrant location, fire department access, private fire protection systems, and location of any special hazards.

We should be aware of some of the guides that help identify the special hazards, particularly hazardous materials that will affect fire fighting operations, that may be found during prefire planning and during fire prevention inspections.

Marking Systems for Hazardous Materials

The primary and most common method of marking hazardous materials is with symbols, written labels, and one- or two-word warning placards. The time for fire fighters and occupants to be aware of these markings and their warnings is *now,* before a fire starts or an accident happens. Therefore, prefire planning and inspections once again are seen to be of primary importance in

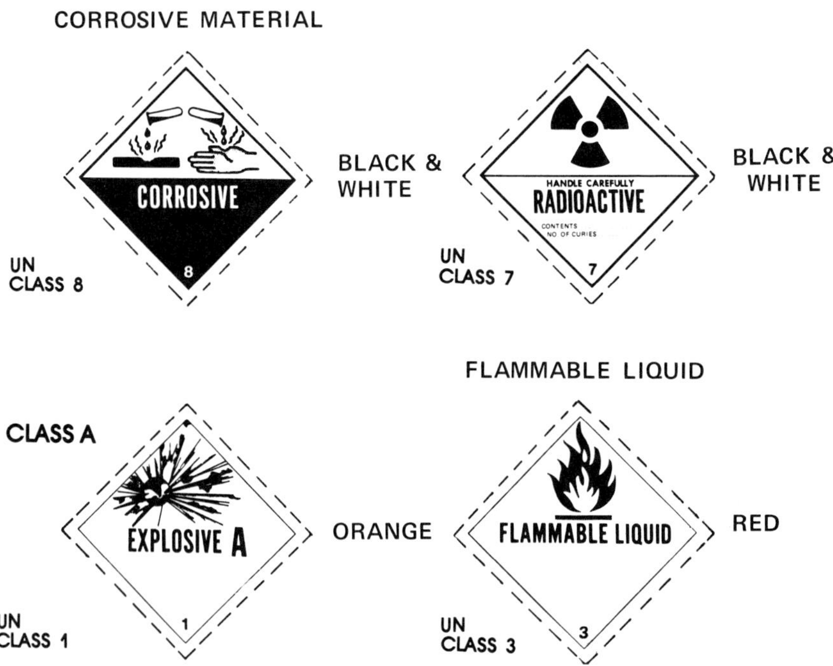

Fig. 11.9. Examples of U.S. Department of Transportation labels used for Hazardous materials that are shipped interstate. (U.S. Department of Transportation.)

fire protection. A dark smoke- and heat-filled room, with flames licking about and things going boom is no place to be trying to read labels.

The most common marking system seen is that of the Department of Transportation (Fig. 11.9). These labels are required for all hazardous materials transported interstate. In addition,

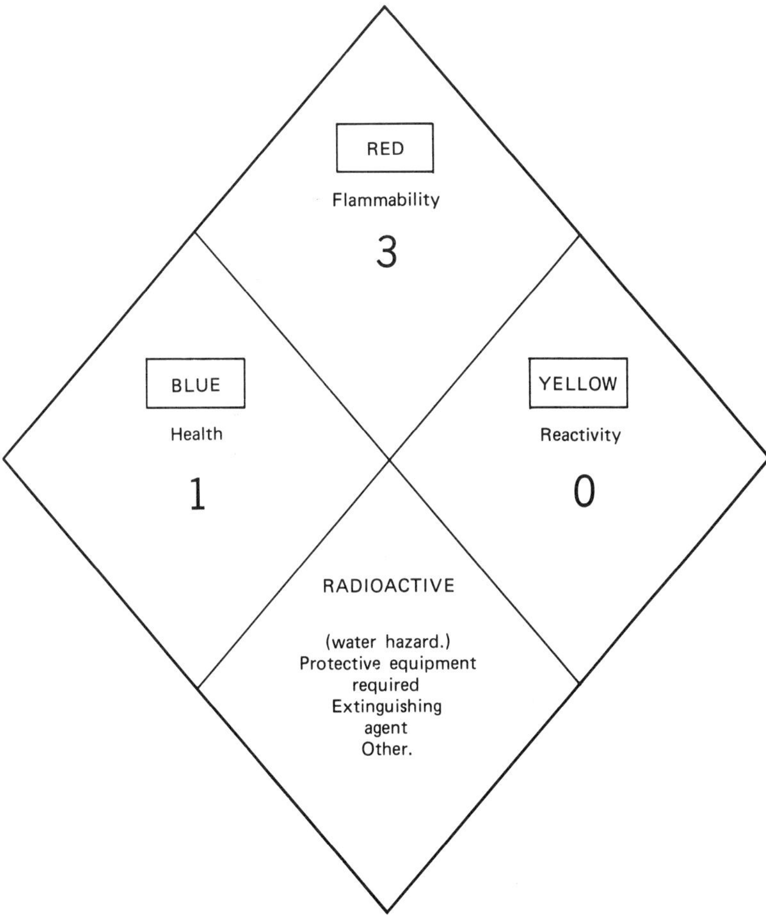

Fig. 11.10. An example of the National Fire Protection Association's marking system. The marking shown here is for acetone. The bottom square is used to indicate any special hazards such as water reactivity, radiation hazard, or other problems. (Based on the 704 Identification System, NFPA.)

Fig. 11.11 NFPA System of Identifying Hazards of Materials

Identification of Health Hazard Color Code: BLUE		Identification of Flammability Color Code: RED		Identification of Reactivity (Stability) Color Code: YELLOW	
Type of Possible Injury		Susceptibility of Materials to Burning		Susceptibility to Release of Energy	
Signal		Signal		Signal	
4	Materials which on very short exposure could cause death or major residual injury even though prompt medical treatment were given.	4	Materials which will rapidly or completely vaporize at atmospheric pressure and normal ambient temperature, or which are readily dispersed in air and which will burn readily.	4	Materials which are readily capable of detonation or of explosive decomposition or reaction at normal temperatures and pressures.
3	Materials which on short exposure could cause serious temporary or residual injury even though prompt medical treatment were given.	3	Liquids and solids that can be ignited under almost all ambient temperature conditions.	3	Materials which are capable of detonation or explosive reaction but require a strong initiating source or which must be heated under confinement before initiation or which react explosively with water.

3	Materials which on intense or continued exposure could cause temporary incapacitation or possible residual injury unless prompt medical treatment is given.	2	Materials that must be moderately heated or exposed to relatively high ambient temperatures before ignition can occur.	2	Materials which are normally unstable and readily undergo violent chemical change but do not detonate. Also materials which may react violently with water or which may form potentially explosive mixtures with water.
1	Materials which on exposure would cause irritation but only minor residual injury even if no treatment is given.	1	Materials that must be preheated before ignition can occur.	1	Materials which are normally stable, but which can become unstable at elevated temperatures and pressures or which may react with water with some release of energy but not violently.
0	Materials which on exposure under fire conditions would offer no hazard beyond that of ordinary combustible material.	0	Materials that will not burn.	0	Materials which are normally stable, even under fire exposure conditions, and which are not reactive with water.

Fig. 11.11. The NFPA numbering system used to indicate the relative hazard within each hazard (i.e., health, flammability, reactivity). (Reprinted with permission from the NFPA Fire Protection Handbook, 14th edition, Copyright 1976, National Fire Protection Association, Boston, MA.)

similar labeling, with some variations, is required if the materials are carried by rail, ship or barge, or by air freight. The drawback of these labels is that they only indicate the major hazard of the material, with markings such as DANGEROUS, FLAMMABLE, CORROSIVE, and other similar one/word warnings in bold type.

One system that does give more information at a glance (when compared to the Department of Transportation's marking system) is that recommended by the National Fire Protection Association. Unfortunately, it is not widely used, nor is it required to be by law. The NFPA marking system uses colors and numbers which indicate, to those who know how to interpret the marking, full information about the hazards of a material. Examples of the NFPA marking system are shown in Figs 11.10 and 11.11. Note that the health hazard, fire hazard, and reactivity hazard (caused by chemical change such as decomposition) are noted in the label.

Other warnings of the hazards of materials are provided by manufacturer's associations who have their own codes or systems. An example is that used by the Compressed Gas Association to identify the contents of compressed gas containers (e.g., green for oxygen). However, these systems are not always reliable—they vary in actual use. Experience, research, and a good course in hazardous materials are a must for fire fighters who have to deal with the rapidly changing technology and materials of today. One never knows what will be encountered in warehouse and manufacturing plant fires and those in fires involving transportation such as trucking, rail freight, and aircraft cargo. The key to safety is continuous learning, preparation, and efficient planning.

SUMMARY

Our man-made environment provides both shelter and hazards. However, good construction can aid in minimizing many of the built-in hazards and in the confinement of fire. In order to understand how materials in building construction affect fire protection, we must be able to identify terms pertaining to construction.

Occupancy and type of use indicate contents and how a

building is or will be used. The type of use and occupancy affect life safety and the potential for fire spread, and are the keystones used to determine construction requirements.

Type of construction refers to the actual materials used in construction and the ability of these materials to withstand fire penetration and fire spread. Types of construction are divided into categories beginning with fire-resistive and ending with unprotected combustible. Within these general categories, building codes often use roman numerals or numbers for subdivisions (e.g., Type I, Type III, and others). Each type of construction requires specific materials and ratings in hours for walls, ceilings, roofs, and other structural components.

The ratings for construction components are established by testing against specific criteria established by testing laboratories. The number of hours or fractions of hours construction assemblies resist fire penetration is given as the fire-resistive rating (e.g., 2-hour wall). The use of rated materials can determine how well a building will withstand fire and aid in reducing ultimate fire loss.

Interior finish and contents are two more major factors that determine the spread or control of an interior fire. These two will also affect exiting and life safety, if not controlled. In many structures, the interior finish must meet a specific flame-spread rating and decorative materials must be treated to resist flame spread.

The contents of a building will affect fire spread, and, if poor storage methods are allowed to exist, may even block access and egress to a building. The possibility of fire may be increased by the storage of highly hazardous or highly reactive materials with other contents.

A fire department that uses prefire planning with a strong prevention program can alleviate many hazards within buildings and at the same time prepare for a possible fire. Fire suppression crews, under the direction of their company officers, should make on-site surveys of target and other hazards, then make plans for the best attack in the event of fire. By the use of written reports, survey maps, and sketches, all the important features of a particular building that are relevant to fire protection can be noted, and a plan of action can be decided upon before a fire occurs.

In order to determine special protection needs and take precautions, members of a fire department and others concerned with fire protection must be able to identify the hazards of materials. Basically two marking systems are used for indicating the dangers of hazardous materials. The most widely used marking system is the one required by the Department of Transportation for all materials shipped interstate by any method. The second system, established by the National Fire Protection Association, is better, but not widely used; it is not required unless it is demanded by the local fire jurisdiction. Both systems indicate the special hazards of the materials labeled.

REVIEW QUESTIONS

1. Why is the occupancy and type of use important to fire protection and building construction?
2. Is there such a thing as "fireproof" construction? Explain your answer.
3. How does type of construction affect fire protection and life safety?
4. What is another designation for Type I construction? Type III? Type V?
5. Define, in your own words, fire-resistive rating. How does it differ from flame-spread rating?
6. How are fire-resistive ratings determined?
7. Why is the control of interior finish important to life safety?
8. Define decorative material.
9. List some of the hazards that may affect exitways.
10. Should prefire planning and fire prevention inspections be conducted separately? Why?
11. Define the purpose(s) of prefire planning.
12. What is a target hazard?
13. List some of the items that should be noted in a prefire plan report.
14. What are the limitations of the Department of Transportation hazardous materials marking system? When or where are the markings required?

15. What hazards are noted in the NFPA marking system?
16. Are the colors of compressed gas cylinders meaningful? Why?

REFERENCES

American Iron and Steel Institute, *Fire Protection Through Modern Building Codes*, 4th ed., New York, NY, 1971.

William K. Bare, *Fundamentals of Fire Prevention*, chapters 5 and 8, John Wiley & Sons, New York, NY, 1977.

National Fire Protection Association, *Fire Protection Handbook*, Section 6, Chapters 1-10; Section 18, Chapter 1, Boston, MA, 1976.

International Conference of Building Officials, *Uniform Building Code*, Whittier, CA, 1973.

chapter 12

12 FIRE PROTECTION EQUIPMENT AND SYSTEMS, PUBLIC AND PRIVATE

In the previous chapter we discussed how building construction may aid in fire protection. Structures also have built-in fire protection systems, such as standpipes, automatic fire sprinklers, and alarm and detection systems. In addition, extinguishers are utilized for "first aid fire protection" and are perhaps the most common equipment of any of the private fire protection systems. In this chapter we will discuss these private systems and another system, not necessarily a part of a building but essential to fire protection—the fire hydrant system. The hydrant system may be privately or publicly owned; however, the design and functions of this system remain essentially the same.

At the end of this chapter we will have completed an

overview of private fire protection systems and learned how they work and how they are utilized by the fire department. The objectives of this chapter are:

1. To identify common fire protection and alarm systems.
2. To identify the major components of each type of system discussed.
3. To identify the common extinguishing agents used in fire protection.
4. To identify how private fire protection systems can aid fire suppression efforts and life safety.

New Terms

Dry barrel hydrant
Wet barrel hydrant
"Making a hydrant"
Automatic fire sprinklers
Underground
Overhead
Riser
Sprinkler head
Fire department connection
Wet pipe system
Dry pipe system
Preaction system
Deluge system
Cyclic system
Wet standpipe
Dry standpipe
Extinguisher classification
Extinguisher rating
Extinguishing agent
Special hazard extinguishing system
Heat detector
Rate of rise
Flame detector
Smoke detector
Ionization and products of combustion detector

We will start our survey of private fire protection systems with the most common system seen throughout our cities and counties—the hydrant system.

HYDRANTS

We see fire hydrants every day, but probably do not stop to think of them as anything other than something we don't park in front of or as something we really wouldn't want in our front yard. After all, a hydrant may be functional, but for the most part it isn't something that would win art awards. In fact, the fire department would like them to be as visible as possible, although many people would like to cover them with ivy or surround them with screens to reduce their visibility, especially when the hydrant sits in the front yard. Hopefully, for fire protection, the fire department can keep them visible and accessible by constant hydrant inspection. Be that as it may, what is a fire hydrant?

Basically a fire hydrant is a large chunk of cast metal containing a valve to turn the water supply on and off and having one or more threaded outlets for fire hose, which are protected by metal caps. These hydrants are the only readily visible part of a large water supply system buried beneath the ground. To be of value for fire protection, these hydrants must supply large amounts of water whenever needed by the fire department.

There are two basic types of hydrants. The first is utilized in areas subject to extreme cold and is referred to as a "dry barrel" hydrant. In this type of hydrant there is no water in the visible portion of the hydrant until the valve is turned. The second type of hydrant is the "wet barrel," which has water generally present in the hydrant itself; however, the water does not flow out of the hose outlets until the valve stem is turned on. To prevent tampering, the valve stem on both hydrants, usually located at the top of the hydrant, requires a hydrant wrench, which is carried on the fire department apparatus. The caps on the hose outlets must be secure, yet removable when the outlet is needed; their removal is usually accomplished by the use of the spanner wrench carried by every fire fighter. The hydrant outlet caps are important; if only one outlet is going to be used, the other caps remain in place to prevent loss of water, which would reduce the water supply to the hose line in use. During a fire, a fire fighter is assigned to connect the hose and operate or open the hydrant on signal; this is referred to as "making a hydrant." In large fire departments where there is enough manpower, the operation and use of the hydrant may be the special job of a fire fighter referred to as a hydrantman.

Now that we know a bit about the hydrant and its use, let's delve further into the portion of the system that is usually hidden from view.

The diagram in Fig. 12.1 will give you some idea of the hidden portion of a hydrant system. It is a hypothetical system, but it does show a water supply, valving, mains, and other components of a system. It should be noted that where possible all the mains (underground pipe) are interconnected to avoid dead-end mains, which reduce the available gallons per minute and pressure, when more than one hydrant is opened on the main.

The location of the hydrants in a given system is determined by the types of occupancies within the area served; usually a fire department officer and a water company official jointly decide the required size of the water main and the spacing, location, and type of hydrants. The inset in Fig. 12.1 shows the common shutoff valve locations employed in a section of a water main supplying

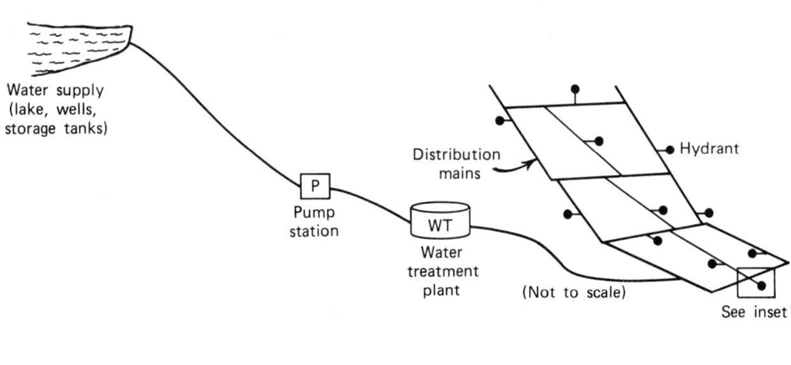

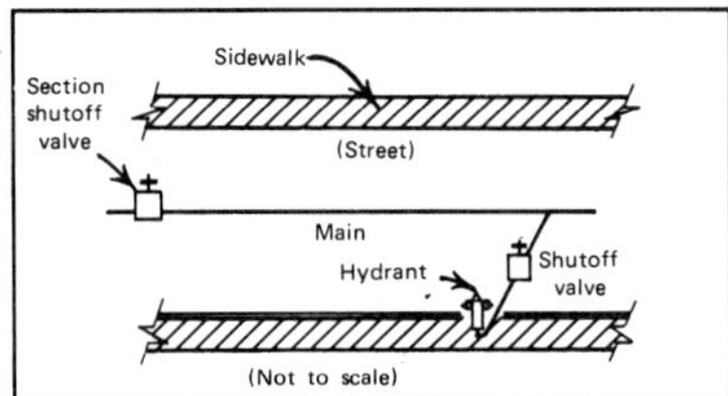

Fig. 12.1. A hypothetical hydrant system. Note the inset which shows the control valves within the system.

hydrants. These shutoffs are used in the event of a break in the main, or if a hydrant is damaged or broken and must be replaced. It is obviously important to the fire department that these shutoffs remain open and that the hydrants function properly. To insure the viability of hydrants, they are periodically inspected and opened (flushed) to check that the valves in the street are open, and that there are no obstructions in the mains. This process may be handled by the water company or department or by the fire department or, even better, by both of them working together. Through working together, a good relationship is created between the fire and water authorities; and joint efforts also help the fire department personnel to become familiar with all the aspects of the water supply system, including hydrant location and amount of water available at any given hydrant.

There are sources of water for a fire department other than hydrants, and Fig. 12.2 illustrates some examples. Of course, a single pumper would not utilize all of these sources at the same time.

Another system that employs water and is similar to a hydrant system in many respects is the automatic fire sprinkler system, more commonly known as "sprinklers."

AUTOMATIC FIRE SPRINKLER SYSTEMS

"Instant fire fighter" might be an appropriate name for an automatic fire sprinkler system, considering how it operates, and how well, when properly designed, installed, and maintained, it does its job. It is one of the few fire protection systems that can offset construction defects in protecting against fire. In addition, it is a major factor in protecting life, due to its ability to either completely extinguish a fire in its earliest stages or hold a fire in check until fire fighters arrive. Sounds good, but what does it consist of and how does it work?

Quite simply, an automatic fire sprinkler system is a series of interconnected pipes of different diameters which provide a connection from a water supply to each individual sprinkler head installed in the system (Fig. 12.3). A typical sprinkler system has four major segments, which are:

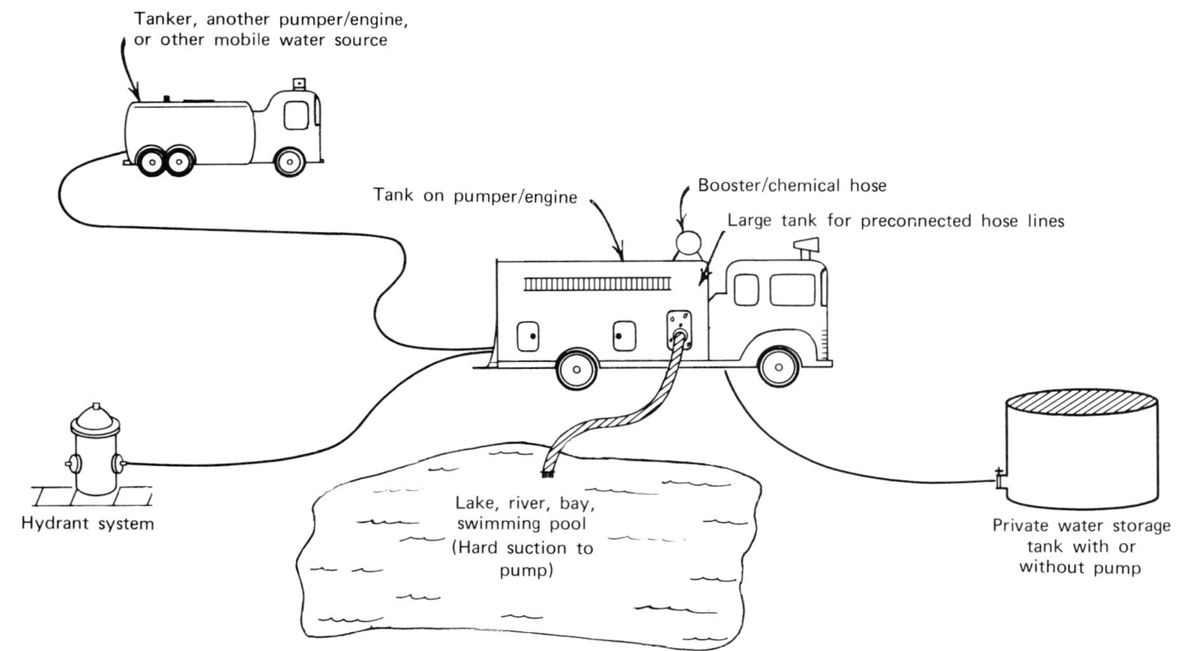

Fig. 12.2. Examples of water supply sources for fire fighting. Not all the illustrated sources would be used at the same time.

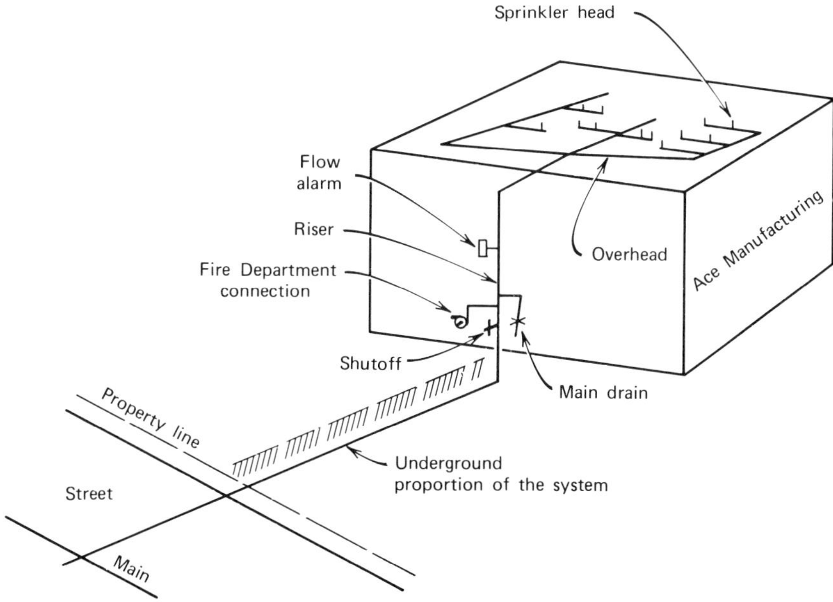

Fig. 12.3. A hypothetical automatic fire sprinkler system which shows the main components and their locations in relation to one another.

Underground
Riser
Overhead
Sprinkler heads

Each of the first three segments consists of piping located as indicated by the descriptive name. That is, the underground is the portion of piping placed underground, connecting the water supply (public water system, private storage tank) with the riser. The riser is the vertical pipe which transfers the water from the underground to the overhead. Depending on the design of the system, the fire department connection, control valves, and other necessary equipment will be found on the riser, or, in some instances, on the underground. The basic purpose of the fire department connection is to allow the fire department to connect hose lines to the system and supplement or provide the water supply to the overhead.

Control devices function just as the name implies; they control the water supply to the system, and may be opened and closed to facilitate repair or replacement of any part of the overhead. Alarm devices are also used to indicate when water flows in the system, either through sprinkler head actuation or a break in the system.

The overhead portion of a sprinkler system consists of piping of various diameters supported on metal hangers, which hold the overhead piping in place. In a simple system all the overhead piping is supplied by a single riser, with the various overhead mains branching out from the riser and suspended from the ceiling(s). In some buildings the overhead piping is hidden from view, with only the sprinkler heads showing through at intervals in the ceiling covering.

The sprinkler heads themselves are attached to the branch mains of the overhead piping, and the proper spacing between the sprinkler heads must be determined by building contents, construction, and physical layout of the floors to be protected. Each sprinkler head, with the exception of the "open" type, reacts individually to heat (Fig. 12.4); that is, they do not all operate at the same time. The "closed" type sprinkler heads are manufactured to operate at specific temperatures (e.g., 165° F, 212° F) and are of different designs (e.g., pendant, upright, exposure, sidewall) that give varied spray patterns. In addition, entire sprinkler systems can be designed to meet various needs and may be of different types.

Ready for action, Viking sprinklers have the water opening securely capped. The cap is held in place with two opposing lever arms, which are in turn held by a Viking ball bearing fusible link.

In case of fire the solder in the link fuses. Instantly the links separate on free-rolling ball bearings. The lever arms, having an offset center, are thrown clear by the spring pressure of the frame. The water cap is thrown out by water pressure.

Through the calibrated orifice water strikes the deflector. The design of the deflector then distributes the water in the uniform pattern typical of all Viking sprinklers.

Fig. 12.4. A typical "closed" sprinkler head in operation. (Courtesy The Viking Corp.)

Types of Systems

Essentially five types of systems are available for use today. These types are:

Wet pipe
Dry pipe
Preaction
Deluge
Cyclic

Wet pipe. The most common type of sprinkler system is the wet pipe system, which contains water in the sprinkler system pipes at all times and uses sprinkler heads of the closed type. Each head has water under pressure behind it, and the water is released when the heat melts the link between the deflector and the base of the sprinkler head. The other types of sprinkler system are essentially of the same design as the wet pipe, with the following major differences:

Dry pipe. The water supply is held in check at the riser by compressed air in the overhead. When a sprinkler head (''closed'' head) opens, the air pressure is released, which allows water to flow to the heads.

Preaction. Again, water is kept out of the overhead until needed. The sprinkler heads are of the closed type, and heat detectors are used to detect a fire. When the detector senses fire (actually a rise in air temperature), an electrical signal is sent to the control valves on the riser and the water is allowed to flow into the overhead.

Deluge. Basically this system is the same as the preaction, except that all the sprinkler heads are of the ''open'' type. Therefore, once the heat detector signals for water, all the heads in the system operate at once, giving a large volume of water over the entire area covered by the system.

Cyclic. This system is the same as a wet pipe system, with the exception that heat detectors are added along with a device that will automatically shut the water off as the heat is reduced. If the heat increases, the system operates again automatically. This type of design reduces the likelihood of water damage but is not in widespread use at this time.

All the types of systems discussed can be designed for use with additives, such as foam, plus water. For example, aircraft hanger fire protection may utilize a foam and water deluge system.

There is much more to a sprinkler system design than we have covered here, but we now have covered the essentials of automatic fire sprinkler systems.

The system we discuss next is also a combination of water and piping: the standpipe.

STANDPIPES

Once again we find ourselves describing a series of interconnected pipes that supply fire fighting water to specific locations. However, instead of a hydrant or a sprinkler head distributing the water, we find hose connections and shutoff valves located along the length of the pipe. Standpipes are generally found in multistoried buildings and in structures that have large open areas, such as warehouses and buildings with public assembly occupancies. Depending on the design and type of standpipe system, there may be water in the pipe at all times (wet standpipe), or there may be an arrangement of valves that allows water to flow through the pipe when a valve is opened (combination standpipe), or the pipe has no permanent source of water supply other than the water provided by the fire department as needed at the time of a fire (dry standpipe). The first and last types of standpipes are the more common, so let's examine these two a bit further.

Wet Standpipe

Wet standpipes are those that you see in buildings, installed in cabinets and/or racks with hose and nozzle attached. The hose is connected to a shutoff valve and is usually 1½ inches in diameter. The source of water for the wet standpipe is the public water main and/or a private tank and pump. The standpipe is filled with water at all times; when water is needed for a fire the occupants have only to unreel the hose and turn on the valve. Although considered as a hose to be used primarily by the building's occupants, the fire department may also use the wet standpipe; when they do, they usually attach their own hose and nozzle to the standpipe outlet at the fire scene.

Dry Standpipe

Unlike the wet standpipe, the dry standpipe, in most instances, is used only by the fire department. No hoses are provided at outlets (Fig. 12.5) which are usually located in protected stairwells of

Fig. 12.5. A typical dry standpipe outlet.

multistoried buildings; hoses must be brought in by the fire department. The purpose of the dry standpipe is to reduce the amount of hose and the time required to place water on a fire in the upper floors of a building. Without the dry standpipe, fire fighters would have to carry and lay hose from the pumper all the way to the fire floor, and thus be unable to deliver quick fire attack.

The water for a dry standpipe is provided via a fire department connection (similar to the one used for automatic fire sprinkler systems) at the time of need. The hose required is transported by fire fighters to the fire floor and connected to the dry standpipe outlet at or near the fire floor; the valve is then opened, and fire fighting begins. The minimum diameter of piping for a dry standpipe is 4 inches, and the standpipe has a minimum of one 2½-inch hose outlet at each floor, plus a multiple outlet located on the roof.

From standpipes and other water systems, we now move to the most common of all private fire protection equipment: extinguishers.

EXTINGUISHERS

Extinguishers are everywhere—they are the first aid kit of private fire protection. Extinguishers come in all sizes and shapes and with all sorts of extinguishing agents inside, yet the operation and purpose of all extinguishers remains basically the same.

The purpose of a fire extinguisher is to supply a readily available, specified source for fire fighting. However, those who may be called on to use an extinguisher must be trained or the effectiveness of the extingusher will be lessened, if not totally negated. To be effective, an extinguisher must be used quickly, that is, while the fire is still small; and the extinguishing agent must be of the proper type for the kind of fuel involved.

In order to determine if an extinguisher will be effective on a given fuel, all one has to do is look on the label for one or more of the following alphabetical letters, which indicate the classification of the extinguisher:

Classification	*Fuels*
A	Ordinary combustible fuels (paper, wood, etc.)
B	Flammable liquids (gasoline, fuel oil, kerosene)
C	*Electrical equipment (items where live electrical current is present)
D	Combustible metals (magnesium, sodium, etc);

The extinguisher label, depending on the extinguishing agent used, may have one or more of the above letter ratings. For example:

Classification	*Extinguishing Agent*
ABC	Multipurpose powder
BC	Carbon dioxide, dry chemical
AB	Foam
A	Water, soda-acid solution, antifreeze solution
D	Special powders

The label on an extinguisher will also tell you how, in a step-by-step explanation, to operate the extinguisher in order to discharge the extinguishing agent properly. The basic steps for putting an extinguisher into effective operation are as follows:

1. Remove the extinguisher from its location and take to the fire scene.
2. Set the extinguisher down and grasp the nozzle or horn in one hand; then activate the extinguisher as directed on the label (e.g., bump, invert, remove pin and squeeze the handle).

*The use of water on these materials would create an electrical shock hazard.

3. Direct the extinguishing agent at the base of the flames until the fire is out or no more extinguishing agent is left.

It is important that fire departments provide training in extinguisher use for businesses and citizens, in order to promote prompt and effective action to extinguish fires when they are small; this will help in attaining fire loss reduction within a community. A single extinguisher, used quickly, can prevent a large-loss fire or at least may hold a fire in check until the fire department arrives.

Our next topic is automatic fire protection systems for special hazards which, in essence, are extensions of built-in extinguishers. If we keep in mind our knowledge of extinguishers and extinguishing agents and add some piping, nozzles, and control devices, we will have a better understanding of the essentials of a special hazard extinguishing system.

SPECIAL HAZARD EXTINGUISHING SYSTEMS

Special hazard extinguishing systems are designed and installed to provide an extingishing agent over a particularly hazardous operation, or storage area. For example, these systems have been used for restaurant cooking facilities (grease hazard), dip tanks for paint solvents (flammable liquid hazard), and over areas where water would be unsuitable as in computer rooms (electrical hazard). In each instance, the system of piping, nozzles, detectors, and capacity of the extinguishing agent storage container must be engineered for proper application and distribution of the extinguishing agent. A system that could be installed over a flammable liquid tank is shown in Fig. 12.6. As can be readily seen, we have extended the use and effectiveness of our basic extinguisher. The extinguishing system may be designed to use dry chemicals, carbon dioxide, or halogenated gas as extinguishing agent. The design will depend on the characteristics of the particular extinguishing agent being used.

The basic components of a special hazard extinguishing system include:

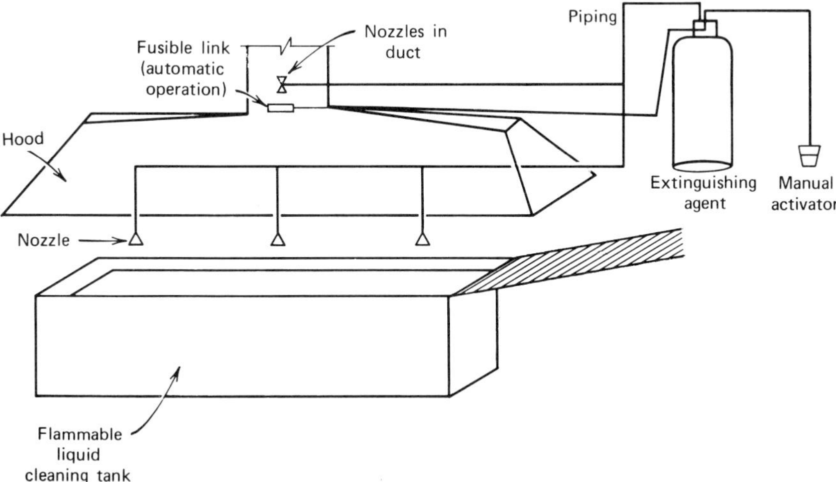

Fig. 12.6. A hypothetical special hazard system; only the main components are visible.

Component	*Function*
Container(s)	To hold the extinguishing agent until needed
Piping	To distribute the agent completely to all nozzles
Nozzles	To distribute the agent completely and evenly over the special hazard being protected
Detectors (e.g., heat, fusible link)	To provide automatic extinguishing action when needed
Pull box, switch	To permit manual activation of the system
Alarm bell, gong, horn	To provide a preactivation alarm in those systems having extinguishing agents that may create a life safety hazard (e.g., a carbon dioxide system)

Let's now complete our survey of private fire protection equipment and systems with those systems that provide alarms and detection.

ALARM AND DETECTION SYSTEMS

We have mentioned some types of alarm and detection equipment (heat dectors, alarm bells, pull boxes) and their purposes, in our discussion of both special hazard extinguishing systems and automatic fire sprinkler systems. We already know that a detector can be used to activate extinguishing systems, and we also know that it may provide prewarning for special hazard systems. In addition to these functions, an alarm or detection system may:

- Warn of tampering with fire protection devices (e.g., a tamper switch can send an alarm if a control valve on a sprinkler system is shut off or moved)
- Warn occupants to evacuate a building (e.g., a school fire alarm system)
- Warn of possible hazardous conditions (smoke detection), and operate building equipment such as fire doors and ventilating equipment
- Notify the fire department to respond to a location (Note: the private fire alarm must be tied into the fire department alarm system to serve this function)

To perform the above functions a wide variety of system designs and components are used, but their essential sensors remain the detectors of various types and pull stations (boxes).

Detectors vary in design and operation and therefore deserve a bit of exploration. It should be noted that each type of detector reacts to a certain aspect of a stage of a fire. Some react more quickly than others, and each has its place in a system design. To help you identify the basic kinds of detectors, we will list them from the slowest reaction time to the fastest and indicate their basic operation.

Type of Detector	*Condition Detected*
Heat	Heat; either fixed temperature or rate-of-rise or both
Flame	Light movement (in cycles per minute) issued by open flame
Smoke	Smoke that is visible to the naked eye
Products of combustion and/or ionization	The products of combustion not visible to the naked eye

As indicated above, the last detector on the list gives the earliest warning of fire—when the smouldering stage has just begun. There are also other types of detectors that serve special functions; for example, explosion detectors help minimize the possibilities of an explosion in special hazard operations involving highly flammable or other materials.

For any detector to be effective it must be placed in the proper location and provide interlocking coverage with other detectors within the system. Therefore, alarm and detection systems are located throughout the area to be protected, but exact placement depends on the type of detector and the configuration of the building. Manual pull stations, where installed, are located so they can be reached quickly in the event of fire.

Alarm and detection systems include devices (bells, horns, gongs, chimes, or prerecorded voice tapes) that warn occupants to evacuate. In addition, a central control panel is necessary to tie all the components together; it may include an annunciator panel which indicates where within a building the warning is coming from.

All of the systems mentioned in this chapter must be installed, maintained, and periodically tested to insure effective operation when needed. This means that fire prevention inspections and prefire planning are necessary to insure that these devices will not fail and that they will function as they were designed to, so that the fire department can utilize each system effectively in the event of an emergency.

SUMMARY

The most common of all fire protection systems is the hydrant system. The hydrants sprout like mushrooms off hidden underground piping throughout most of the country. Although hydrants appear much the same, there are two basic types, designed for two climatic conditions. The dry barrel hydrant is used in areas subject to freezing temperatures, and the wet barrel hydrant for moderate climes. Each hydrant system is composed of a network of pipes which transport water from a source to individual hydrants on the line. Control valves allow isolation and repair of any section of the system without the need to shut down the total system. The location and number of hydrants in any one system is determined by the water supply and needs of the area. Where possible, the locations of hydrants and water main sizes should be determined jointly by the fire department and the water system management.

Another water system similar to the hydrant is the automatic fire sprinkler system. Here again we find a series of interconnected pipes which are used to transport water from a source to outlets (sprinkler heads) within a given area. The controls for a wet pipe system can be located either on the underground or the riser; in others (e.g., dry pipe, preaction), the controls are usually located on the riser.

The sprinkler heads themselves (except the open type) are activated individually by the heat produced by a fire. Some sprinkler systems use a combination of open sprinkler heads and heat detectors to achieve the desired fire protection. Sprinkler systems can be designed to provide additional extinguishing agents along with water for special hazard areas.

Standpipe systems are used to supply water for fire fighting in the upper stories of buildings and for those occupancies where they are required for code compliance, such as in public assembly buildings. There are two basic types of standpipe systems: wet and dry. The wet standpipe is normally considered as being for the use of the building's occupants and the dry standpipe for the use of the fire department.

Fire extinguishers are probably the most common private fire fighting tool found in the man-made environment. Extinguishers are required by fire codes in almost all occupancies, and come in many sizes and with various extinguishing agents. Each extinguisher is classified by the type of fuel(s) that it is effective on, and is rated on a relative scale as to the potential amount of fire it will handle when used promptly and correctly. Fire departments and private fire brigades should train occupants of buildings in the use of extinguishers and other private fire protection equipment (e.g., wet standpipes).

Special hazard extinguishing systems are extensions of the basic extinguisher. Utilizing the correct amounts and locations of precisely engineered piping, nozzles, and detection and activation devices, these systems can protect specific special hazard areas.

Alarm and detection systems both warn and accomplish several functions beyond warning. The capabilities of any system depend on its design and intended function. The eyes of a detection system are those devices that detect and provide a signal for alarm. Each device reacts to a different phase of fire (e.g., smoke, heat, flame), and they can be used in conjunction with one another or with other system components such as automatic fire sprinklers.

All of the systems noted in this chapter must be inspected, tested, and maintained periodically in order to insure effective operation and use. Therefore, fire prevention inspections and prefire planning are a must for insuring proper operation of the systems and maximum fire protection.

REVIEW QUESTIONS

1. What is the difference between the two basic types of hydrants noted in the text?
2. Why are shutoff valves installed on hydrant mains?
3. List some of the factors that dictate the location of hydrants and the size of the water supply main.
4. What is the purpose of flushing a hydrant?
5. List the components of an automatic fire sprinkler system.
6. In what type of sprinkler system do all the sprinkler heads provide water at the same time?

7. In what type(s) of system(s) are heat detectors used in conjunction with sprinkler heads? What kind of sprinkler heads are used?
8. Do all closed sprinkler heads that are available for installation operate at the same temperature? Cite the reason for your answer.
9. Define and list the components of a wet stand pipe system.
10. Define and list the components of a dry standpipe system.
11. What are the purposes of standpipe systems?
12. A class ABC extinguisher will be effective on what type(s) of fuel?
13. List the important points of extinguisher operation.
14. Cite examples of uses for special hazard extinguishing systems.
15. Outline the components and functions of a special hazard extinguishing system.
16. List the kinds of extinguishing agents that may be used in a special hazard extinguishing system.
17. What are the possible functions of an alarm and detection system?
18. List the kinds of detectors that are used in an alarm and detection system and explain the phases of fire that each kind will react to.
19. If you were to choose a home fire detector, which one would you choose? Why?
20. List some of the warning devices used in an alarm system. Cite examples of occupancies in which you have seen alarm systems.

REFERENCES

National Fire Protection Association, *Fire Protection Handbook,* 14th ed., Sections 11–16, Boston, MA, 1976.

William K. Bare, *Fundamentals of Fire Prevention,* Chapters 6 and 7, John Wiley & Sons, New York, NY, 1977.

chapter 13

13 COMBAT: TACTICS AND STRATEGY

Up to this point, all of our chapters have discussed the basics of fire science, fire prevention, and fire protection. In this chapter we will describe the operations of a fire department when an emergency occurs. In the truest sense of the word, this is combat: the pitting of people and machines against a natural force that will spare nothing in its path. On the fire scene there is no time for discussion or meetings. Everyone must know what to do, how to do it, and make the least number of mistakes in achieving the objective—saving life and property.

If you have ever witnessed fire fighting operations at a fire scene, you may have felt as though all hell had broken loose: engines roar, hose flies, fire fighters run, and equipment is scattered here and there. Fire fighters are seen chopping holes in perfectly good roofs, breaking windows, and tearing holes in walls, seemingly bent on creating more destruction than the fire is creating. However, it is all by plan, and for very good reasons. Once you are aware of the tactics and strategy involved in fire fighting, the chaos is seen as a system. This system is the subject of the present chapter. In addition to discussing the tactics involved in fighting building (structure) fires, we will discuss wildland and transportation fire tactics, which will show how the basic tactics are adapted to different kinds of fires.

At the end of this chapter you will be able to:

1. Define fire fighting strategy.
2. Define fire fighting tactics and identify the basic components.
3. Identify the differences between fire tactics involved in building, wildland, and transportation fires.

New Terms

Strategy	*Extinguishment*
Tactics	*Salvage*
Size-up	*Overhaul*
Rescue	*Mop-up/patrol*
Explosives	*Slow burning fuels*
Confinement	*Fast burning fuels*
Ventilation	*Backdraft and smoke explosion*

STRATEGY

Fire fighting strategy is similar to that utilized in military operations, and thus the word "combat" again appears meaningful in reference to fire fighting. The strategy of fire fighting involves the best possible management of resources on the fire scene to achieve the objective of extinguishing the fire with minimum loss of life and property. The resources used by the fire ground commander include men and equipment, built-in fire protection systems (e.g., fire doors, sprinkler systems, standpipes), and the elements (e.g., wind, humidity, heat) and are organized to produce the maximum effect in the shortest possible time.

Tactics, on the other hand, consist of a series of steps used to gain smaller specific objectives on the way to achieving the overall objective.

TACTICS

We will discuss those utilized in the operational sequence involved in fighting a building fire. Later we will see how these same tactics are adapted to fight transportation and wildland fires.

The process utilized in resolving any problem, including fire fighting, amounts to assembling information in a series of steps that lead to a decision and action. We might say that we use tactics every day in order to accomplish our personal objectives or resolve our problems. In fire fighting, as with any other undertaking, we must first define our problem and find out what is known as opposed to what the possiblilites are. In the instance of fire fighting this step is known as size-up.

Size-up (Evaluation)

Size-up is the keystone to the balance of the tactical steps and is what eventually determines the strategy used to achieve fire fighting objectives. However, size-up does not begin upon arrival at a fire; it must begin at the time the alarm is received. (Now prefire plan information becomes important).

The person who bears the initial responsibility for size-up is the company officer of the first-due engine company. When the alarm is received in the station, his initial size-up begins as he is given the report of the type of fire, its location, and other information by the alarm operator (Chapter 6). At that point, the company officer begins to assemble the known facts.

The facts that he considers, as should every member of the company in his own mind, relate to items such as:

- Kind of emergency reported (smoke, actual fire, explosion)
- Location of the emergency (which determines the quickest route to the location and, if prefire planning has been

done, calls to mind mental images of hydrant locations and type of construction.

- Type of occupancy (which helps determine possible rescue problems)

Other relevant facts are time of day, weather, and other knowns such as manpower and apparatus available from other stations. The time of day and weather are considered because they both affect apparatus response, fire behavior, and rescue, among other things. For example, the time of day will govern the kinds of traffic problems that may be encountered and may possibly increase the amount of time needed for response. Thus, a response to an alarm at 5:00 may be slowed due to heavy traffic as people are getting off work, whereas a response to an alarm received at 2:00 may be quicker due to little or no traffic. In addition, time of day will possibly affect the need for rescue; at night, for the most part, people are indoors and asleep, and their chances for being trapped by fire are greater. Hospitals and similar occupancies usually operate with a reduced staff at night, which means that a greater number of fire fighters would be needed to rescue and evacuate patients.

At this point, we must remember that all of this portion of size-up occurs as a mental process, after the alarm is received, while leaving the station, and enroute to the fire scene; this process will continue until fire ground operations are concluded. In other words, mental size-up will go on and be adjusted by observations which, taken all together, make up the final strategy.

Once the fire department arrives on the scene other basic tactical steps take place to achieve the fire fighting objectives. One of the first considerations at a fire or any other emergency is life safety; thus, *rescue* takes priority and is the next focus of tactics to be discussed.

Rescue

The need for rescue and the operations necessary for rescue are determined at the fire scene, based on information gathered from the occupants and/or bystanders and by observation (e.g., someone hanging out of a window screaming for help). The company officer must then determine how to effect rescue and if extinguishing operations must begin in order to make the rescue possible or if rescue alone must proceed first. If the rescue requires most or all of the manpower at the scene or if special equipment is needed (e.g., rescue squad and/or paramedical aid), the company officer must request, via radio, further aid to insure positive action without allowing the fire to escape his control; the additional help may require a change or adaptation of tactics.

The next series of tactical steps actually often occur at the same time and are not readily distinguishable as they happen. In essence, four tactics work together and are aimed at the outer edges of the fire to box it into a smaller and smaller area and to prevent further spread; this eventually leads to complete extinguishment. For this reason we will group these tactical steps of *exposure, confinement, ventilation,* and *extinguishment* into one section, which we will call checkmate.

Checkmate

Just as chess requires certain moves to achieve checkmate and the end of further play, the tactical steps of explosure, confinement, ventilation, and extinguishment are used, as necessary, to stop a fire. Let's examine each of these steps.

Exposures are those areas of a building or adjacent buildings or structures that may become involved in fire as a result of the heat or actual flame produced by the main fire. As we know from earlier chapters, fire can be spread without direct flame contact via convection, conduction, and radiation. Therefore, the company officer must insure that no further spread of fire will occur by taking steps to reduce the exposure of adjoining buildings or areas. He may accomplish this control by having master stream appliances set up between buildings to act as heat absorbing curtains to prevent additional fire through exposure. In addition, hose streams may be directed on smoke columns and on adjoining combustible roofs to prevent flying embers from setting fires. In short, the company officer must use all his knowledge of the subjects we have covered in this book to protect exposures and confine the fire.

Confining a fire requires knowledge of building construction, hydraulics and hose streams, fire behavior, and much more. In this tactic (confinement) a knowledgeable fire officer will use the features of a building and its construction to aid him in his efforts to confine a fire to as small an area as possible. As we noted earlier, the proper use of fire doors, fire-resistive construction features, and any built-in private fire protection systems (standpipes, sprinklers) will make confinement an easier task. Another factor that can aid in confining a fire and reducing damage is proper ventilation.

Ventilation consists of the removal of toxic fire gases, superheated air, and smoke (unburnt fuel) from a building involved in fire. Ventilation is the responsibility, where manpower permits, of the truck company officer and crew. (Note: Rescue may be the primary responsibility of the truck company in a large department.) Several methods can be used to ventilate a building; these range from opening doors and windows to use of smoke ejectors and forcible methods such as cutting holes in roofs and attics above the fire. The value of proper ventilation is that it removes the smoke, increases the visibility within a building, and makes finding the origin of the fire easier. As proper ventilation also removes the superheated air, it makes the atmosphere within a building bearable for fire fighters and reduces the chance of fire spread by lowering interior air temperature. Last but not least, ventilation helps reduce possible damage to the contents of a building and the building proper from smoke and heat. However, if ventilation is ill timed or not properly executed, it can increase the fire spread and can cause an explosion, known as backdraft or smoke explosion.

This type of explosion occurs when a high concentration of unburnt fuel particles (smoke) is present in an oxygen-depleted atmosphere. If ventilation occurs without proper precautions, such as having hose streams in place, the cool oxygen-rich air from the outside of the building rushes in and ultrarapid oxidation (explosion) occurs.

Extinguishment is the last tactical step that we consider as part of checkmate. This is an obvious step in fire fighting; however, it can be done well or poorly, depending on the training and size-up capabilities of the fire department and its officers. The proper use and selection of hose streams, knowledge of building construction, and judicious application of water by fire fighters, all combine to achieve rapid and complete extinguishment with a minimal loss from fire or water damage.

The key to rapid extinguishment of a fire is the ability to see the actual fire, rather than a red glow in a smoke-filled area and then to make the environment tolerable for fire fighters so they may enter and apply either straight streams or fog to effect extinguishment with a minimum of water damage. As we know from our discussion, proper ventilation can aid in extinguishment; therefore, a fire attack may combine these two tactical steps so they occur at nearly the same time.

Effective extinguishment activities take into consideration the kinds of contents present in order to protect the fire fighters from unnecessary risk. For example, if the fire were in a magnesium casting plant, and the metal itself were part of the fuel burning, the situation would call for modification of the usual extinguishing efforts. Use of water would create a more serious fire (magnesium will react violently with water) and endanger fire fighters; therefore, the company officer would alter the normal extinguishment activities to suit the type of contents involved. Again the importance of prefire planning becomes evident. Other hazardous materials that require changes in operations are flammable liquids, compressed gases, and toxic or reactive chemicals. We can see that many bits of information must be part of the company officer's size-up if he is to achieve checkmate.

Having covered the main extinguishing operation, we move to the last two tactical steps which are salvage and overhaul.

Salvage and Overhaul

Overhaul is a term that is used to describe the final, complete extinguishment of a fire to prevent a rekindle. In forest or other wildland fires, this same operation is referred to as ''mop-up.'' In both instances the efforts include checking to make sure that all embers are out and that the property is left in as safe a condition as possible. As we know from our discussions of fire investigation, overhaul must be accomplished carefully in order to prevent the scattering of vital evidence as to the cause of the fire. In both overhaul and salvage operations (which occur at almost the same time) the basic goal is to minimize damage.

Salvage refers to the removal of the remainder of the smoke (via smoke ejectors), removal of water from the interior of a building by the use of vacuums and other methods, and to the use of salvage covers to protect undamaged materials from water damage. Where there is sufficient manpower, salvage operations can be conducted during extinguishment to help protect contents on floors other than the fire floor (in multistoried buildings) from water seeping down from the floors above. Again, the main purpose of these tactics is to save as much property as possible from fire and fire fighting operations.

At this point we conclude our general discussion of fundamental tactics and strategy; as you now see, a definite methodology is used in fire fighting, regardless of how it may look at the fire scene. We now will show how these tactical steps are adapted to meet different situations; for this purpose we will discuss tactics used at wildland and transportation fires.

WILDLAND TACTICS

Whereas a building has built-in devices to aid in the reduction of fire spread (i.e., walls, roofs), an open agricultural or wildland area does not. Therefore, our basic tactics and strategy must be adapted in order so successfully extinguish a fire in open lands. To understand this adaptation, let's look at size-up.

The officer who will go to the scene of a wildland fire must start a mental review of response problems and available manpower and equipment at the time the alarm is received—just as the officer who responds to a building fire. Once on the scene, because of the different type of fire, the size-up must be altered.

First, the officer must try to get a clear view of the total fire—which may not be easy—to determine his method of attack. (This same effort applies, when possible, to sizing up building fires). Wind direction, terrain, type of fuel, and weather, all play determining roles in deciding on the proper method of fire attack. For example, if there is a fire in a level field of grass and a strong wind is blowing, the chances of being able to stop the fire at its "head," or the point at which it is advancing, are small. I have experienced this type of fire, and it burnt faster than we go in our trucks, until it reached a paved road which blocked its further spread and allowed us to gain control by attacking its sides.

Other examples of how terrain and weather will affect fire behavior in wildland fires follow:

- A fire will spread rapidly up a slope or hillside and burn slowly downhill.
- Normally, a fire will die down at night (due to increased humidity, fuel moisture, and less wind) and burn rapidly during the day.
- In a canyon or valley the wind will blow up canyon during the day and down canyon at night.

Fire is also affected by the burning rates of fuels; fuels can generally be classified as fast or slow burning. Fuels considered *fast burning* are dry grasses and grains, dead leaves and needles, dead or dry brush, and small bushes and trees. *Slow burning* fuels are logs, large branches, stumps, and heavy, moist leaves (duff) found on the floor of thick forests.

After the fire officer considers the above factors, his next step is checking or stopping fire spread; again, he is following the same basic tactics outlined in checkmate. In wildland fires, several options may be available to achieve checkmate, but the basic one involves utilizing manpower to physically limit the readily available fuel. This is not to say that hose lines are not used—they are, where feasible; however, this is not always the case. Therefore, manpower and tools, in addition to natural barriers such as roads and rivers, must be used to contain the fire.

In remote or rural areas where water is not readily available, fire fighting utilizes fuel removal to contain a fire and some unusual methods of extingushment. For example, for fuel removal, bulldozers may be used to cut away potential fuel leaving the mineral soil level to aid in containing the fire. In addition, hard work by fire fighters using shovels, rakes, axes, brush hooks, and chain saws to remove fuel from the edges of the fire results in a man-made fire line. Aircraft (e.g., vintage bombers, helicopters) may be used to transport personnel and to carry water and chemicals, such as fire retardants, which are dumped on the edges of the fire to stop the fire's spread. Aircraft, then, becomes a tool that a fire officer may use to extinguish a wildland fire. Other items used for extinguishment range from the prosiac (green braches and wet gunnysacks to beat out flames) to the backpack pumps that contain five gallons of water. An unusual extinguishment method which must be used carefully is the backfire. Backfiring is the setting of a man-made fire, by fire fighters, to burn away fuel and to stop fire spread. Now that we know a bit about how to get a fire under control and the main steps in extinguishment, we move to the next steps: mop-up and patrol.

Mop-up, as indicated earlier, corresponds to overhaul in building fires. Mop-up includes removal of snags (dead and burnt standing trees), extinguishment of smoldering materials, widening and clearing of the initial fire line, and similar activities. Even with a fire of a few acres, mop-up can take as long as a day, or more, to insure that rekindle from hidden smoldering pockets of wood, roots, or other fuels is impossible.

Patrol is the last tactic of wildland fire fighting and begins during the last stages of mop-up. This tactic, requires that fire fighters be assigned to patrol the fire perimeter to watch for any signs of rekindle or pockets of smoldering material. In addition, lookouts may be posted to observe the area surrounding the burnt area for signs of small fires outside the fire perimeter. Only after experienced officers have judged a fire to be completely out and the area safe will the manpower and equipment be able to leave the fire scene. Large fires may require several days of duty, living out of a fire camp established near the fire area. As with building fires, investigation as to the cause of the fire will continue from the beginning of mop-up until the cause(s) is found.

Now we leave wildland fire tactics and discuss the problems of transportation fires.

TRANSPORTATION FIRE TACTICS

The term transportation fire here refers primarily to fires involving truck, air, and rail transportation of hazardous materials. These types of fires may be encountered in any fire jurisdiction, and therefore the tactics they demand deserve special mention. Although the basic tactical steps are adapted to the special circumstances and utilized, we find some variations in the size-up.

Once on the scene of the fire, one of the first and most urgent steps is identifying the cargo involved and its potential life safety hazard, which coincides with rescue of victims at the scene. The company officer may obtain his first information as to the nature of the cargo from placards on a boxcar or truck. A placard may state "FLAMMABLE," "EXPLOSIVES A," "OXIDIZER," or "CORROSIVE." His next step is to determine the exact nature of the cargo, which may not be as easy. For purposes of illustration, we will discuss fire involving a truck carrying explosives.

The company officer, upon arrival, should attempt to speak to the truck driver to ascertain what kinds and the quantities of explosives being carried. Unfortunately, he may not be able to get information from the truck driver because he is injured or not at the scene. I was told by one truck driver who carried explosives, "If this thing ever catches fire, I'm going to jump out of the cab and run like hell." Well, that might solve his problem; however, it would leave fire fighters in a very poor position.

Regardless of how much or how little information is received about the cargo, the company officer's next consideration is the location of the fire in the truck, which will determine his next tactical steps. For example, if the fire is in the cargo area itself, he will have to evacuate the area for at least 2000 feet in all directions and fight the fire from a safe distance with the use of master streams. If the truck involved is a tractor-trailer assembly, and the fire is confined to the engine area, he may decide to separate the two to keep the fire away from the cargo within the trailer. In fighting a fire of this nature, the two main concerns that determine the tactical operations are to keep the fire from reaching the explosives and, if this is not possible, to keep everyone, including fire fighters, well away from danger; thus, the company officer plans his tactics from the standpoint of protecting life and the surrounding areas, and may let the fire go.

As is evident by now, the same basic tactical methods are used regardless of the type of fire; however, they must be be adapted to a certain extent for each fire. Also, it may be noted that tactics and strategy require the full use of all aspects of fire science in order to achieve the goal of protection from fire. All of the topics we have covered in this and preceding chapters contribute to safety and are utilized by those in the field of fire protection.

Now we proceed to our final chapter, which will include information regarding entry into the fire protection field. Employment in this area may be your goal. If it is not, then from this book at the very least you have gained an overview of fire protection and its problems, which will help you to understand both the fire service and the need for improved fire protection in all of our communities.

SUMMARY

The combating of fire requires planning, organization, and the ability to adapt to conditions on the fire ground. Achieving the ultimate goal of fire extinguishment with minimal property loss and no loss of life requires good fire fighting tactical operations and sound strategy.

The basic tactics used for fire suppression begin when the alarm is received and continue throughout the fire fighting until complete mop-up or overhaul has occurred. Size-up or evaluation, the first (and a continuous) step employed by fire officers, begins with review of the known facts. These facts include available manpower, time of day, location of emergency, type of emergency, and similar facts. Upon arrival at the fire scene, size-up continues while the other tactical operations required to meet the emergency are implemented.

Rescue is the next consideration and may be the first physical action taken at the fire scene. The fire officer must determine a course of action which will effect rescue; at times, fire fighting operations may be delayed until rescue operations are completed. However, there are times when an initial fire attack is required before rescue attempts can begin. The next four tactical steps box a fire in and effect extinguishment. The first step is insuring that *exposures* are protected. Once this is accomplished, the fire is contained within the building of origin, and during *confinement* the leading edges of the fire are driven back. During this step, *ventilation* of the building may be required to vent smoke and heat from the fire area. As the fire is reduced in size, eventual *extinguishment* occurs, leaving the *overhaul* and *salvage* operations to be accomplished.

Wildland (forest, grass, grain, brush) fire fighting tactics are basically the same as those employed in building fire fighting tactics, but are adapted to fit the circumstances. Terrain, fuel moisture, weather, and other factors are important aspects of wildland fire size-up. Also the types of fuels (just as the types of contents in a building fire) must be considered, as they influence the type of fire attack to be used. For the most part, rural and wildland fires require the type of attack that consists of fuel removal, owing to the lack of available water for fire fighting. Fuel is removed with tools, manpower, and machinery. After the main extinguishment is accomplished, *mop-up* and *patrol* begin to insure that the fire is completely out.

Transportation fires involving cargos of explosives and other hazardous materials require adaptation of our basic tactics. In some instances, no attempt is made to extinguish the fire because of highly dangerous conditions or potential of the fuel; thus, evacuation and protection of exposures are the only actions that can be taken without undue risk.

REVIEW QUESTIONS

1. How does fire fighting strategy differ from fire fighting tactics?
2. What are the tactical steps involved in fighting building fires?
3. What are some of the conditions that must be considered during size-up or evaluation?
4. How does weather affect fire fighting tactics in building fires? Wildland fires? Transportation fires?
5. Why is rescue one of the first considerations at a fire scene?
6. What are the tactics involved in "checkmate"? Can one or more be accomplished at the same time? Why?
7. What must be considered before ventilation occurs? Why?
8. Why do smoke or backdraft, explosions occur?
9. What is the difference between salvage and overhaul? Mop-up and salvage? Overhaul and patrol?
10. List the effects of terrain on wildland fire tactics.
11. What are the differences between fast and slow burning fuels? List examples of each.

12. What are some of the tools used for wildland fire control and extinguishment?
13. How can a fire officer identify the cargo involved in a truck fire?
14. When should a fire fighting attack be withdrawn in a fire involving a truck transporting explosives?
15. In your opinion, why is evacuation to 2000 feet necessary in a fire involving explosives? Where would you place fire fighters if you were in charge of the fire scene?

REFERENCES

Lloyd Layman, *Fire Fighting Tactics,* National Fire Protection Association, Boston, MA, 1953.

Warren Y. Kimball, *Fire Attack 1,* National Fire Protection Association, Boston, MA, 1966.

———*Fire Attack 2,* National Fire Protection Association, Boston, MA, 1968.

U. S. Forest Service and the California Division of Forestry, *Forest Fire Fighting Fundamentals,* 51992, pp. 11–61.

PART IV

IV
FIRE PROTECTION AND THE FUTURE

chapter 14

14 WHERE NOW?

Now we are coming to the end of our journey through this book and are considering beginning of a larger undertaking—a Fire Service career. This chapter is devoted to helping a potential career develop into a reality. If your are planning a career in fire protection engineering or any other fire protection career that demands a four-year college degree, then your course is dictated by the college of your choice and by time.

However, if you want experience in the field, or more specifically, a budding career in the public Fire Service you may wish to start to work in short order. To help you in the search for a job in a fire department this chapter will deal with where to apply, how to apply, and what to expect while seeking to obtain entrance into a civil service fire protection career. The chapter is divided into how and where to look for job openings, entrance examination preparation, testing procedure, and examination aids. The last section is devoted to a brief discussion of promotional examinations and aids.

Before we get into how and where, let's identify the objectives of this chapter.

At the end of this chapter you will be able to:

1. Identify the steps in a civil service entrance examination procedure.
2. Identify the function of each step of the entrance examination process and its effects on the eventual ranking of an applicant.

3. Identify the basic components within each step of the entrance examination process.
4. Identify the aids that may help an applicant choose a suitable fire department and prepare him to do well in the entrance examinations.
5. Identify the steps in a promotional examination and the aids that will help the applicant to be successful.

New Terms

The new terms used in this chapter are

Eligibility list	*Physical agility test*
Oral board	*Rank and position*

EXPLORATION

I think it is most important to do a little exploration prior to deciding where one wants to apply for a position. For example, at this time moving between fire departments can be a bit costly in terms of lost seniority, vacation time, promotion, and retirement benefits. In other words, it might be wise initially to pick the fire department that offers you the most promotional opportunities and other benefits that meet your needs, rather than just take the first job from any department. As you climb the promotional ladder, it may become increasingly difficult to give up what you have gained on the chance that another fire department may offer you more at a later date.

It is important to remember before choosing a department that very few promotional jobs are opened to persons employed in other fire departments; the traditional approach is promotion from within, except for some chief of department jobs. At this time, the best opportunities for moving between fire departments occur in the chief's ranks, and primarily in the specialist areas, such as fire prevention and fire investigation, and there are some openings in training. Hopefully, standardization and certification programs will become widely accepted, and the traditional hiring practices (promotion held to within a department) will change at all levels within the Fire Service. Then perhaps lateral transfers between departments without loss of accrued benefits will become more available. This is something for the future; right now we are more interested in finding the job that best suits your needs.

Probably the least time-consuming way of exploring the current job market in the Fire Service is via the U.S. Postal Service. Send letters to the personnel departments of those cities that you feel as though you would like to live and work in. In your letter request information about the fire department and benefits, even though no openings may be available at the time. If you are not sure if the fire department is a municipal agency (fire district, county fire department), address your letter of request for job information to the Chief of the Fire Department. Another way of obtaining information regarding benefits, pay, and hours is to go to the nearest public library and consult a copy of the latest *Municipal Yearbook,* published by the International City Management Association, Washington, D.C. This informative resource book will give you the latest information regarding the activities of both Canadian and American municipal governments.

Once you have completed all of your research and have the comparative information in hand, decide which cities and departments to explore further. If possible, visit each city or area that you have a definite interest in and stay for at least one or two days to see how it feels. Staying a while gives you the opportunity to go to at least one or two fire stations of the same department (providing there is more than one station) and talk to the men on duty—fire fighters are friendly and helpful people for the most part, so don't worry about being rebuffed. Ask the men for their estimation of their jobs and the department in general. If you can stay another day, go back to the same stations and ask the same questions of the men on the other shift. The reason I suggest making this second visit is because of human variations. Fire fighters are not all goodness and light, and you may run into a disgruntled fire fighter who gives an unfavorable picture of a department, which may mislead you. By talking to several men you will probably have enough information to make your own decision about the department in question. Using your visits to towns and fire stations, you can narrow down the field of desirable departments. Now the next step is getting a job with the fire department of your choice, and in order to accomplish this you must know about a job opening and take the entrance test.

Usually—unless there are immediate job openings—a personnel department will not take any applications but will take your name and address and notify you when the test you are interested in comes up. However, do not totally rely on a clerk in a personnel office to notify you of an impending opening and test; check about every two months to make sure that you will have ample notification of a job opening. If you are considering federal civil service fire protection jobs, you should contact the local Federal Job Information Center (listed in the phone book under "United States Government") or call the Regional Job Information Center. At your request they will tell you if there is a job announcement in your field (Fig. 14.1), and, if so, they will also send a copy of the announcement and other material. However, finding the job opening is just the first step on the road to finally working for a fire department. Civil service starts with the filling out of an application form.

APPLYING FOR THE JOB

The application blanks for the various governmental agencies vary somewhat, but all require the same basic information. A city government application form is shown in Fig. 14.2, and Fig. 14.3 shows a federal government application form.

There are some hints that can make the filling out of application forms less time-consuming and tedious. The first is that you can take the form to a quiet place, home if possible, and fill it out completely in a slow relaxed manner, making the application look as neat as possible. If you can type, type the form; if not, print or write clearly. If the form cannot be read easily, your chance of a job may be jeopardized.

The second hint is to have a ready-made list of past work experience (supervisor's name, name of firm, address, phone

Announcement No. FR–3–10
Issued: December 10, 1973
Open Until Further Notice
No Written Test

FIREFIGHTER
(Structural, Airfield, General)
GS–4, GS–5

FIRE PROTECTION INSPECTOR
GS–6, GS–7

Limited Employment
Opportunities with Various
Federal Agencies in the
San Francisco Bay Area
Central Coastal Region

SAN FRANCISCO AREA OFFICE
U. S. CIVIL SERVICE COMMISSION

Fig. 14.1. A sample federal job announcement.

*DO NOT USE BLUE INK

DO NOT USE THIS SPACE	APPLICATION FOR EMPLOYMENT
REVIEWED BY ______	PERSONNEL OFFICE
ACCEPTED ______ REJECTED ______	CIVIC CENTER
REASON ______	110 E. MAIN - P.O. BOX 949 LOS GATOS, CALIFORNIA 95030

POSITION APPLIED FOR ______

NAME (PLEASE PRINT) ______ LAST ______ FIRST ______ MIDDLE

ADDRESS ______ STREET ______ CITY ______ STATE & ZIP CODE ______ PHONE NO. ______

S.S. NO ______ DRIVER'S LIC. NO. ______

ARE YOU NOW, OR HAVE YOU EVER BEEN EMPLOYED BY THE TOWN OF LOS GATOS? YES ☐ NO ☐

HAVE YOU EVER BEEN DISCHARGED FROM ANY EMPLOYMENT? YES ☐ NO ☐ (IF YES, GIVE DETAILS ON REVERSE SIDE)

HAVE YOU EVER BEEN CONVICTED OF AN OFFENSE OTHER THAN MINOR TRAFFIC VIOLATIONS? YES ☐ NO ☐ (IF YES, EXPLAIN FULLY ON REVERSE SIDE)

IF YES, GIVE DEPT. ______

(DATES) FROM ______ TO ______

IF RELATED TO ANY TOWN EMPLOYEE, GIVE NAME ______

RELATIONSHIP ______ DEPT. ______

PERSON TO NOTIFY IN CASE OF ACCIDENT ______

ADDRESS ______ PHONE NO. ______

EDUCATION

	CIRCLE HIGHEST GRADE COMPLETED		
GRAMMAR AND HIGH SCHOOL	1 2 3 4 5 6 7 8 9 10 11 12		
COLLEGE OR UNIVERSITY (NAME OF SCHOOLS)	1 2 3 4 5 6 7	DEGREES	NO. OF COLLEGE UNITS COMPLETED
	MAJOR		
BUSINESS, TRADE OR CORRESPONDENCE SCHOOL	COURSES STUDIED		

EMPLOYMENT RECORD

LIST ALL EMPLOYMENT FOR PAST 15 YEARS WITH MOST RECENT EMPLOYMENT FIRST.

EMPLOYMENT DATES	EMPLOYER,S NAME AND ADDRESS	OCCUPATION AND DESCRIPTION OF DUTIES	SALARY RECEIVED	REASON FOR LEAVING
FROM				
TO				
TOTAL TIME				
FROM				
TO				
TOTAL TIME				
FROM				
TO				
TOTAL TIME				
FROM				
TO				
TOTAL TIME				
FROM				
TO				
TOTAL TIME				

(CONTINUED ON REVERSE SIDE)

Fig. 14.2. A typical city or county job application form. (Courtesy Town of Los Gatos, CA.)

IF APPLYING FOR CLERICAL EMPLOYMENT, INDICATE SPECIAL SKILLS YOU HAVE ACQUIRED

☐ SHORTHAND SPEED______W.P.M.
☐ TYPING SPEED______W.P.M.

☐ CALCULATING MACHINES (KIND)
☐ MIMEOGRAPH
☐ DRY COPY MACHINE (KIND)
☐ ADDRESSOGRAPH
☐ P B X

DO YOU HAVE ANY PHYSICAL DEFECTS? ☐ YES ☐ NO. IF YES, EXPLAIN:

USE THIS SPACE TO FURNISH ADDITIONAL INFORMATION, SUCH AS CONTINUATION OF EDUCATION OR EXPERIENCE RECORD. SUMMARIZE ANY ADDITIONAL INFORMATION NECESSARY TO DESCRIBE YOUR FULL QUALIFICATIONS.

CERTIFICATE OF APPLICANT. (READ CAREFULLY BEFORE SIGNING)

I hereby certify that all statements made in this application are true and I agree and understand that any misstatement of material facts herein will cause forfeiture on my part of all eligibility to any employment in the service of the Town of Los Gatos.

SIGNATURE ______________________ DATE______

number, dates of work, and beginning and ending salary per hour or per month). In addition, if you have trouble remembering the dates you graduated from schools or the names of schools you attended, make a list of those also. Having these lists, and keeping them up to date, makes filling out job applications simple and accurate. It also makes filling out forms less tedious because you can copy the required information off your master list, rather than struggle to remember your salary two years ago or who you worked for four years ago and how long you worked there.

PERSONAL QUALIFICATIONS STATEMENT

Budget Bureau
Approved 50-R0387

1a. Kind of position *(job)* you are filing for *(or title of examination)*
b. Announcement number
7. Birth date

Month	Day	Year

8. Social Security Number

c. Options for which you wish to be considered *(if listed in examination announcement)*

d. Primary place(s) you wish to be employed

9. If you are currently on a register of eligibles for appointment to a Federal position, give the name of the examination, the name of the office maintaining the register, the date on your notice of rating, and your rating.

2. Home telephone

Area Code	Number

3. Business telephone

Area Code	Number

4. Name *(Last)* *(First)* *(Middle)* *(Maiden, if any)* ☐ Mr. ☐ Miss ☐ Mrs.

10. Lowest pay or grade you will accept

PAY		GRADE
$ per	OR	

5. Number and street, R.D., or Post Office box number

6. City State ZIP Code

11. Are you willing to travel? *(Check one)*

NO	SOME	OFTEN

12. When will you be available?

13. Will you accept:		YES	NO	(C) Will you accept a job in:	YES	NO
(A) Temporary appointment of	—1 month or less?			—Washington, D.C.?		
	—1 to 4 months?			—any place in the United States?		
	—4 to 12 months?			—outside of the United States?		
(B) Less than full time work?	*(Less than 40 hours per week)*			—only in *(specify)*:		

14. EDUCATION

(A) Did you graduate from high school, or will you graduate within the next nine months?

YES	MONTH/YEAR	NO	HIGHEST GRADE COMPLETED

(B) Name and location *(city and State)* of last high school attended

(C) Name and location *(city, State, and ZIP Code if known)* of college or university. *(If you expect to graduate within 9 months, give MONTH and year you expect degree.)*	Dates attended		Years completed		Credits completed		Type of degree	Year of degree
	From	To	Day	Night	Semester hours	Quarter hours		

(D) Chief undergraduate college subjects	Credits completed		(E) Chief graduate college subjects	Credits completed	
	Semester hours	Quarter hours		Semester hours	Quarter hours

(F) Major field of study at highest level of college work

(G) Other schools or training *(for example, trade, vocational, armed forces, or business)*. Give for each the name and location *(city, State, and ZIP Code if known)* of school, dates attended, subjects studied, certificates, and any other pertinent data.

15. HONORS, AWARDS, AND FELLOWSHIPS RECEIVED

16. FOREIGN LANGUAGES

Enter foreign languages and indicate your knowledge of each by placing "X" in proper columns	Reading			Speaking			Understanding			Writing		
	Excl	Good	Fair	Excl	Good	Fair	Excl	Good	Fair	Excl	Good	Fair

17. Special qualifications and skills *(licenses; skills with machines, patents or inventions; publications—do not submit copies unless requested; public speaking; memberships in professional or scientific societies; typing or shorthand speed; etc.)*

THE FEDERAL GOVERNMENT IS AN EQUAL OPPORTUNITY EMPLOYER

Standard Form 171 (Formerly SF 57)
July 1968 U.S. Civil Service Commission
171-101

Fig. 14.3. A typical federal job application form.

After you have completed the application, review it to make sure it is completely filled out. Don't leave spaces that do not apply blank; put in the letters "N.A." (not applicable) so the reviewer knows that the information was not omitted by error. The last check should be to make sure you have signed your name. It is amazing how many people go to great lengths to complete their application, only to fail to sign it. An unsigned paper is not too impressive to the reviewer, and the application is usually reviewed by several people, as we will find out later in this chapter. A list of some important reminders regarding the completion of application forms is shown in List 14.4. Once the application is returned to the personnel office, it undergoes a long process that will eventually bring a job.

Fig. 14.4. Important points to remember when filling out job applications. (Courtesy Donna Marquis Bare.)

Give full and complete information—street numbers, phone numbers, first names where possible, and titles of former supervisors.

Avoid lengthy job descriptions—they take up space and most interviewers won't read them.

Print or *type* on all applications, except when directed otherwise; use a *pen* unless a pencil is specified. Print clearly and avoid crossing out—*think* it out before putting it down on paper.

Be accurate regarding dates of employment—they are easily checked. *Do not* exaggerate about duties, position, or salary.

When a question on an application does not apply, print "N.A." or "none" in the space provided; this indicates to the reviewer that the question was not overlooked.

Do not turn in the application without reading it over to check for errors or possible ommissions.

Generally, all applications are reviewed for completeness and checked to see if experience, education, and physical condition meet the minimum job qualifications. Listed references may be checked immediately or later when the final selection is about to be made. Therefore, it is important that all the information in your application be accurate, that is, not falsified in any manner. (Note: falsification on an application is generally grounds for dismissal, even after you have gotten the job.) If, based on the information you give, you are found to meet the minimum qualifications, you will be notified of the time, date, and place to take the written entrance examination.

THE WRITTEN EXAMINATION

The written examination is the second step in the selection process. The test is used to screen applicants and reduce the number of applicants to those who do best on the test. Most written entrance examinations are basically general intelligence tests covering such areas as:

Mechanical Aptitude
Reading Comprehension
Mathematics
Judgment
Basic Aptitude for Fire Fighting Work

There is usually a time limit but very few applicants require the full time allotted to complete the written test. As with any examination, there are restrictions (special pencils, seating arrangements, time limits on certain sections of the test), and each jurisdiction may have its own special test section.

To help you study for the written test, you may desire to use one or more of several publications. Various publishing houses sell these under the general headings of Civil Service Entrance Examinations or Fire Department Entrance Examinations. I have never used these aids for any test that I took (either entrance or promotional), so I cannot comment on their usefulness. To make your own decision, visit the public library and look in the reference section for copies of these aids. (At the end of the chapter, publishers of these examination aids are listed.)

Your acceptance for the next step in the selection process is based upon your written test score. Usually a score of 70 percent or more is required to qualify for the physical agility and other tests in the selection process. (Note: Some variation occurs between agencies; that is, some have the physical agility immediately after the written test, before the scores on the written examinations are known. Always check to make sure that you will be prepared, by asking the personnel department to explain the sequence in the selection process.) Those who achieve less than the minimum qualifying score on the written test are not allowed to compete further. However, they may compete at a later date in the event of a new job opening, and the selection process will begin again for them. So the written test reduces the number of candidates competing for the job. At this time it should be pointed out that anyone who has taken the written examination does have the right to review his test with the supervision of the personnel department at a specified time and date. Now, let's assume that a satisfactory test score has been achieved and move on to the next test, the physical agility.

PHYSICAL AGILITY TEST

The physical agility test is important for both the candidate and the fire department. Because fire fighting is a strenuous and hazardous job requiring stamina, agility, and working at heights under stress, this test will hopefully disclose which candidates are capable of performing physical tasks under some pressure without endangering themselves or others. The test doesn't require the candidate to be a superman nor does it require that he be extremely fit, but it does require that the candidate be in good health, without serious physical impairment, and in reasonably good shape. Those candidates who are overweight (have a poor weight-to-height ratio for their age and body structure) and those who are unaccustomed to moderate physical activity will probably have some difficulty in meeting the minimum standards for performance.

A typical physical agility test includes some sort of timed running (e.g., running a measured course within a given time), an obstacle course of some sort to test coordination, a test of strength (perhaps lifting a set amount of weight a given number of times within a given time period). The test activities vary; however, most of the tests are designed to measure stamina, coordination, and strength. Perhaps special efforts to train for an agility test will pay off, but if you are reasonably fit and physically active, the test should not be too great an obstacle, providing you can conquer any inherent nervousness at performing in front of others.

I mention "in front of others" because most agility tests are given to candidates in a group, so other candidates will be watching your performance as well as the persons rating your performance—these are usually fire department officers. Each candidate's performance is rated on a grading sheet, and at the end of the agility test these grading sheets are used to augment the written test scores. Those candidates who failed to meet the minimum acceptable performance requirements are then dropped from further competition. For those who qualify it is on to the next hurdle, which is supplied by the Oral Examination Board.

THE ORAL EXAMINATION BOARD

For most people the prospect of facing an oral examination board sets the butterflies to roaming freely in the stomach, causes beads of perspiration to glisten on their foreheads, and starts their minds to wondering if another career wouldn't be better. In reality, I have never been a participant in an oral board whose members did not understand the feelings of the candidates facing them, nor have I ever faced an oral board that I felt treated me unfairly. The people who compose an oral board are human, and have been on both sides of the table and realize what the candidate's fears are. So don't fear an oral board examination but enjoy it for what it is—your chance to express yourself as an individual. You will probably always be nervous regardless of how many times you appear before an oral board in a Fire Service career, but the amount of nervousness you feel may be reduced if you have some idea of the purpose of an oral board and what kinds of questions may be asked. The oral board exam result is an important factor in the overall final score of a candidate, and how the results are used can help put the oral examination in perspective. So, for all these reasons let's explore this exam.

If you have been successful in the written and physical agility tests, the oral is the last major hurdle, other than the final physical examination. Each candidate is notified of the time, date, and location for the oral examination. By this time the total number of candidates that began with the written test has been reduced—usually by half—so those attending the oral make up a much smaller group. Each candidate is allotted time for an individual interview (i.e., examination) with the oral board. (Therefore each candidate arrives at a specified time for his own appointment.) The time alloted amounts to about 20 to 30 minutes. It is wise to arrive at least five minutes prior to the scheduled time, to insure that you will not be late for your interview. Being late for a scheduled oral interview should be avoided. If for some reason the scheduled time cannot be met (e.g., illness in the family, accident), the candidate should notify the personnel department as soon as possible and make arrangements for a different time. Courtesy is the name of the game.

At the appointed time, the candidate is ushered into the room and introduced to the oral board members. The composition of an oral board varies, but there are usually at least two to three fire department officers (from departments other than the one that has the job opening), a member representing the personnel department, and perhaps a member of the Civil Service Commission, if there is one. Normally, the person from the personnel department oversees the introductions, explains a bit of each candidate's background before the interview, shows the board members the candidate's application and test scores, and collects the individual rating sheets on the candidate from each board member.

Each board member will usually ask the candidate a certain number of standard questions relating to the reasons for wanting the job, why the candidate feels he is qualified for the job, and other questions. After each response, the board members rate the candidate and his responses as to poise, ability to communicate, and other matters, without discussing the candidate with the other board members. The purpose of the board is to interview each candidate and determine suitability for the Fire Service while allowing the candidate to ask questions and express himself to the best of his ability. For the candidate, it is the time that he can "sell" himself, in a positive sense, and show his desire for the job, as well as his ability to communicate. However, a word of caution: don't overdo the enthusiasm or meander in a word maze when responding to the questions asked by the board members. The questions should be answered completely and honestly, but without a lot of verbal garbage and false gusto. In essence, be yourself and respond naturally. If you do so, your chance of success will probably be better than if you try to act the way you think the board wants you to act.

Usually at the end of the oral interview, each candidate is given the opportunity to express any important information about himself that was not brought out during the questioning from the oral board. At the conclusion of the interview it is time for courtesy, a farewell to the board members, then out the door. Now you begin to wonder how well you did. After you leave the oral board, work on your interview continues. Each board member evaluates your responses and his overall impressions. Then they review the next candidate's application to prepare for his interview. At the end of all the interviews, the board members are usually requested to rank each candidate from the best to the last. They usually base rank according to who they would hire if it were their decision. This unofficial ranking is not the actual ranking used, but it is the first evaluation. The final ranking of the candidates is within the province of the personnel department, and is the result of several factors. This brings us to our next consideration: the eligibility list.

THE ELIGIBILITY LIST

The position, or rank, of each candidate on the eligibility list is determined by the personnel department. The list is based upon a combination of written test scores and oral board scores (usually given in percentages). Other factors are considered, but these tests are the basic factors. Normally, the test scores are combined to give a total score; then the total scores for all the applicants are compared. The candidate with the highest total score holds the rank or first position on the list, the next highest score is number two, and so on. Thus, if there are ten final candidates, the candidate with the lowest total score would be ranked number ten on the eligibility list.

There are variations on this method of scoring, such as giving more weight (in percent) to the written test score and less to the oral, or vice versa. Regardless of the system used, both scores are important, and a candidate can pull up a low written score by having attained a high oral test score. However, it is difficult to top someone who has a high score in both the written and the oral.

The candidate is usually notified by either card or letter, of his written score, oral score, combined total score, and his position on the list. He may also be told how long the eligibility list is valid (usually one year); that is, candidates on the list will be eligible for hiring until the list expires. Well, so far, so good, but how do you get a job after all this?

HIRING

To answer the last question we turn to the chief of the fire department and examine his options regarding the listed candidates and the process he may use to determine who is finally hired. Since all testing is time consuming and costly, the tests are usually held only when there are actual job openings. So, for illustration, we will imagine that hiring is done soon after the eligibility list is confirmed. Let's also assume that there are three openings for probationary fire fighter and that there are ten men on the eligibility list.

Given the above conditions, the chief of the department reviews the eligibility list and usually has the choice of taking any of the top three to five men listed (i.e., those candidates ranked one through five). His next step may be to call the candidates chosen and ask if they are still interested in the job and make arrangements to have them stop by for a chat. Based on these individual interviews, the chief may decide which of those called he wants to hire. They may not necessarily be ranked one through three on the list. The chief then notifies the personnel department of his choices, and the three are contacted to take the physical examination (paid for by the city or fire department); at this time fingerprinting and a records check are done by the police department. If all goes well and the proper forms are filled out, the now probationary fire fighters are told when to report for work, the uniform requirements, rules and regulations, and other pertinent information. The usefulness of finding out pertinent information regarding a new job can be illustrated by my own experience.

Prior to my getting a job with the fire department, I had done many things, including fighting brush and grass fires for the California Division of Forestry. In Forestry the meals were provided, and we had the luxury of a cook. Because of this experience, I failed to ask, and no one informed me, that in this city fire department, everyone brought their own food and cooked it at the fire station. My first day on the new job went well in the morning, but then lunch time came around. Everyone reached into their cupboards, bags, or other containers and brought forth their individual lunches. Everyone, that is, except yours truly. Well, not wanting to show I had made a mistake, I managed to find things to do, hoping no one would notice my red face and hungry look. Finally, after everyone had finished their lunch, the captain came into the kitchen where I was sitting and asked me if I had brought anything for lunch, and you know my answer: "No." Fortunately, for my stomach, the captain generously offered me a can of tuna, and that was my lunch that day. While I was eating the tuna, the captain was shaking his head and he asked me if I was sure no one had told me about bringing my food. So, the moral of the story is make make sure you find out all the pertinent information that you can *before* your first day. It may save some embarrassment. That first day begins a long period of training and a six month or one year probationary period.

Now that we have filled the three jobs in our imaginary example, let's determine what happens to those people that did not make the final list. If someone fails to make the eligibility list the first time, it does not bar him from taking the examination the next time it's given after the list expires. However, they must begin again and go through the written, physical agility, and oral tests and the rest of the hiring process. With this we conclude our exploration of entrance examinations. However we should note that the sequence discussed varies with the jurisdiction (federal, state, county, city) and primarily applies to civil service fire departments. Entry into volunteer departments and other branches of fire protection may be different, and more specific information can be gotten from the particular fire protection agency that you are interested in. Before we close our chapter, let's briefly discuss promotional examinations.

PROMOTIONAL EXAMINATIONS

The steps utilized in Fire Service promotional systems are essentially the same as those described in our discussion on entrance examinations; that is, filling out an application for a position, meeting the minimum requirements (usually time in the Fire Service, fire fighting experience, and education), a written and oral or just an oral board without a written test, and ranking on the eligibility list. Usually there is no physical agility test or a physical examination. For promotional examinations, as for entrance examinations, there are published study aids that may help. The questions on both the written (if given or required) and the oral require specific professional fire fighting and Fire Service background information and experience.

The best way to be prepared for a promotional examination is by constant study, rather than a hurried cram 5 minutes after a promotional opening is announced. To keep in readiness for a promotional examination, the professional fire fighter can make a habit of reading professional journals (e.g., *Fire Engineering, Fire Journal*), attending seminars and workshops (state and professional groups), and taking community college classes.

For those fire fighters who want additional study aids, there are publications that prepare readers for specific promotional examinations, such as those for captain, battalion chief, and others. In addition, self-study guides, fire science texts, and home study courses are published. Some of these items are available at local libraries or can be purchased either through a bookstore or directly from the publisher. (See the reference section of this chapter for specific publishers and materials.)

Whatever materials are available, it is still the individual and his desire to be a professional that determine who achieves promotion and helps the Fire Service to improve in quality of service.

To those of you who are the future professionals in the nation's Fire Service, I wish success.

SUMMARY

Exploration and research are an essential part of finding a satisfying career in any field. Perhaps research is particularly important, in some respects, before entering the public Fire Service. At this time moving between fire departments without loss of seniority, vacation time, and other benefits is a bit difficult. For the most part, openings to people outside a particular fire department are limited to chief of department positions and jobs in some special areas, such as prevention, investigation, and training. Therefore, consider and research carefully before you decide on a particular fire department. For the greatest return on time spent, use the mail and the local library to conduct preliminary research on particular fire departments. When you have narrowed the field down to the departments you want, it is time to apply for a job.

The testing procedure for entrance into the fire department (civil service) essentially begins with the filing of a job application. If you have prepared a personal history sheet for your reference, then job applications can be filled out quickly and accurately. Once the application is in the hands of the personnel department and you have met the minimum qualifications, the next step occurs.

The basic civil service entrance requirements for fire fighter follow a format which includes a written test, physical agility, and oral board examinations, and eventually an eligibility list. Each segment of the process acts as a screen, and the end result is, hopefully, a list of the best candidates for the job.

Hiring is based upon the eligibility list and is primarily in the hands of the chief of the fire department and the personnel office. Once hired, the candidate still faces a probationary period of six months or one year before he becomes a permanent employee.

Promotional examinations are essentially conducted in the same manner as the entrance examination. There are variations in promotional requirements, such as no written test and/or no physical agility test. However, preparation for both promotional and entrance examinations is important. Study aids and other educational opportunities, within a college and in seminars, are available. For any test, the way to prepare is not by cramming a few days before the test but by constant upgrading one's education and knowledge.

REVIEW QUESTIONS

1. List the reasons why you should explore several fire departments before deciding upon one particular department.
2. List the places and resources that are available to you for collecting information about particular fire departments. Identify each resource and what it offers in the way of information.
3. How does a list of your past experiences (i.e., jobs) help you in applying for a position in the Fire Service?
4. List the key steps required to fill out a job application.
5. What is the first step in the civil service examination process? What function or purpose does it perform?
6. Why is a physical agility test used in civil service for entry level fire fighter? What kinds of tasks should you be prepared for in a physical agility test?
7. What function(s) does an oral board serve in the testing process? What can it do for you?
8. List some of the preparations you should be sure to make before an oral board examination and explain why they are important.
9. What is an eligibility list? How is it used in the context of Fire Service entrance examinations?
10. If you are ranked number three on an eligibility list, but there is only one opening available can you still get the job? Why?
11. Once you have been hired, what are some of the facts you should find out before you report to work for the first day?

12. Are there any differences between the entrance examination process and the promotional examinations in civil service? Explain your answer.

13. List and explain some of the steps you can take to prepare for a promotional examination.

14. What aids are available to help you study for a promotional or an entrance examination? List them.

REFERENCES

Harry Walter Koch, *Fireman Entrance Examinations,* Ken Book Publications, 1968.

Arco Publishing Company, Inc., New York, publishes the following promotional study aids: "Fire Administration and Technology," "Battalion Chief," "Lieutenant," "Captain".

Davis Publishing Company, Santa Cruz, Ca., "Cram Courses" (for promotional examinations) and Self-study Guides and Tests.

INDEX